Progressive Energy Policy

Series Editors
Caroline Kuzemko
The University of Warwick
Coventry, UK

Catherine Mitchell
University of Exeter
Penryn, UK

Andreas Goldthau
Royal Holloway University of London
Egham, UK

Alain Nadaï
International Research Centre on Environment
and Development (CIRED-CNRS)
Nogent-sur-Marne, France

Shunsuke Managi
Kyushu University
Fukuoka, Fukuoka, Japan

Progressive Energy Policy is a new series that seeks to be pivotal in nature and improve our understanding of the role of energy policy within processes of sustainable, secure and equitable energy transformations. The series brings together authors from a variety of academic disciplines, as well as geographic locations, to reveal in greater detail the complexities and possibilities of governing for change in energy systems. Each title in this series will communicate to academic as well as policymaking audiences key research findings designed to develop understandings of energy transformations but also about the role of policy in facilitating and supporting innovative change. Individual titles will often be theoretically informed but will always be firmly evidence-based seeking to link theory and policy to outcomes and changing practices. Progressive Energy Policy is focussed on whole energy systems not stand alone issues; inter-connections within and between systems; and on analyses that moves beyond description to evaluate and unpack energy governance systems and decisions.

More information about this series at
http://www.palgrave.com/gp/series/15052

Oscar Fitch-Roy · Jenny Fairbrass

Negotiating the EU's 2030 Climate and Energy Framework

Agendas, Ideas and European Interest Groups

Oscar Fitch-Roy
Energy Policy Group
University of Exeter
Penryn, UK

Jenny Fairbrass
Norwich Business School
University of East Anglia
Norwich, UK

Progressive Energy Policy
ISBN 978-3-319-90947-9   ISBN 978-3-319-90948-6   (eBook)
https://doi.org/10.1007/978-3-319-90948-6

Library of Congress Control Number: 2018940739

Cover illustration: © Melisa Hasan

Printed on acid-free paper

This Palgrave Pivot imprint is published by the registered company Springer International Publishing AG part of Springer Nature
The registered company address is: Gewerbestrasse 11, 6330 Cham, Switzerland

PREFACE

Guidebooks recommend travellers visit Brussels in the spring or autumn, known as the 'shoulder seasons'. The weather is cooler than in high summer, the crowds less frenetic and hotel prices more moderate. Although, there is a downside: In common with most of North West Europe, Brussels in autumn can spend days or weeks at a time stewing under a leaden blanket of oppressive low cloud.

On one such cool, overcast Thursday morning in autumn 2014, the European Council met at 175 rue de la Loi. The monumental pink granite building, named for sixteenth-century Flemish philosopher, Justus Lipsius, is the venue for the quarterly summits of the heads of state and government of the EU's member states.

At this particular summit, on the 23rd of October, the 28 politicians were due to take a decision with implications for Europe's energy system, the environment and the global climate. The decision was set to shape the priorities for EU's climate and energy policy between 2020 and 2030. The European Council was working from proposals made by the EU's executive, the European Commission, which proposed that several 'targets' be set for 2030. Recommendations had been made for the level and nature of the targets, backed up by hundreds of pages of analysis.

Like any big decision, the 'EU 2030' framework was riven by trade-offs and compromises. With climate change accepted as a real and epoch-defining challenge and transforming Europe's dirty energy systems into sustainable ones proving extremely complex, the stakes were high. Some countries wanted to move faster than others. There were disagreements over the proper role for the EU in promoting specific clean technologies. The strongly held differences of opinion about the right way ahead meant that the potential for disagreement around the huge oval table was palpable.

And then, in the face of what many describe as the *"greatest challenge of our time"*, Europe blinked.

The consensus that emerged from the European Council that day represented a significant change in course from the one set just seven years earlier by the 2020 targets. The overall target for emissions reduction was the lowest that could be said to be commensurate with the EU's goal of deep decarbonisation by 2050, itself informed by the desire to limit global temperature increase to 2 °C. The targets for renewable energy and energy efficiency were both low, and would not carry any legal weight. 'Politics' was seen to have trumped 'climate'.

But many of the critical choices that determined the outcome had not been taken by the Council at all. They had been taken, in fact, in the months and years leading up to the meeting and were contained in the European Commission's proposals, published earlier in 2014. By the time the Prime Ministers, Presidents and Chancellors took their allocated seats that morning, the '2030 framework' and the ideas it represented had been debated, discussed and raked over by scores of experts and analysts in Brussels and elsewhere—relatively few of whom held official positions in the EU.

In the course of developing their proposals, the European Commission had consulted national and European interest groups. Those same interests had prepared and published proposals of their own. They had undertaken modelling exercises and formed coalitions. They had plotted, planned, persuaded and sought at every turn to make their mark on the options that faced Europe's leaders that dreary Thursday in October.

So, while the grand politics of the summit table played a defining role in the 2030 climate and energy framework, so did the intense activity of a small community of interest groups, lobbyists and campaigners, living and working within a few kilometres of the Justus Lipsius building. This book tells that story.

Penryn, UK   Oscar Fitch-Roy
Norwich, UK   Jenny Fairbrass

ACKNOWLEDGEMENTS

The authors are grateful to the numerous friends and colleagues on whom we rely for inspiration, advice and encouragement. In particular, the members of Exeter's Energy Policy Group, ably led by Catherine Mitchell, have proved themselves as generous as they are wise. The frank and open discussions at ECPR 2018 in Oslo with, among others, Sebastian Oberthür, Elin Boasson, Kacper Szulecki, Tor Inderberg and Inga Ydersbond were invaluable. This work would not have been possible without the support of our mutual friend and colleague, David Benson.

We would also like to acknowledge the financial support provided by the University Association for Contemporary European Studies (UACES) that has funded the Collaborative Research Network (CRN) on 'European Energy' that Jenny Fairbrass has run with co-convenors, Anna Herranz-Surrallés and Israel Solorio, since 2015. The UACES funding enabled the first UACES CRN on European Energy, which took place at the University of East Anglia in June 2015. That workshop provided the catalyst for the collaboration between the authors of this volume. Finally, we thank all of the many stakeholders and experts in Brussels who contributed their time and insights with no promise of reward or favour.

The UK Engineering and Physical Sciences Research Council (EPSRC) supported this research under grant 1402887.

CONTENTS

ABBREVIATIONS

BUSINESSEUROPE	Confederation of European Business
COP	Conference of the Parties (to the UNFCCC)
DG Clima	European Commission Directorate-general (DG) for Climate Action
DG Energy	European Commission Directorate-general for Energy
ECF	European Climate Foundation
ECSC	European Coal and Steel Community
(E)NGO	(Environmental) Non-Governmental Organisation
EU	European Union
EU-ETS	European Union Emissions Trading System
Eurelectric	The Union of the Electricity Industry
EuroACE	European Alliance of Companies for Energy Efficiency in Buildings
FoETS	Friends of ETS
GDP	Gross Domestic Product
GHG	Greenhouse Gases
IETA	International Emissions Trading Association
IPCC	Intergovernmental Panel on Climate Change
MEP	Member of the European Parliament
MSA	Multiple Stream Approach
PRIMES	Price-Induced Market Equilibrium System (model)
RES	Renewable Energy Sources
SECGEN (SG)	Secretariat General
UK	United Kingdom
UNFCCC	United Nations Framework Convention on Climate Change
WWF	World Wide Fund For Nature

LIST OF FIGURES

CHAPTER 1

Introduction and Context

Abstract This chapter surveys the historical development of climate and energy policy in the EU and the attendant scholarly attention paid to environmental and climate policy and politics. The significance of the 2030 framework for the future direction of EU climate mitigation efforts is set out, the authors arguing that the policy represents a distinct shift towards technology neutrality. Following an overview of the literature on EU interest groups and socio-technical transitions, the chapter concludes by identifying the impact of interest groups on the policy agenda as the focus of the study.

Keywords European Union · Climate policy · Energy policy
Socio-technical transitions · Interest groups · Agenda-setting

1.1 Introduction

The purpose of this book is twofold. Firstly and most fundamentally, it is an account of the role played by interest groups in setting the policymaking agenda in the lead up to the decision about the EU's 2030 climate and energy targets. The analysis, and the account that flows from it, is informed by the multiple streams approach (MSA), an established but still evolving theory of the policy process introduced by John Kingdon in 1984. Secondarily, it is a commentary on the implications of our findings

© The Author(s) 2018   1
O. Fitch-Roy and J. Fairbrass, *Negotiating the EU's 2030 Climate and Energy Framework*, Progressive Energy Policy,
https://doi.org/10.1007/978-3-319-90948-6_1

for MSA and for future research. This chapter introduces the policymaking and scholarly context for the rest of the book. It is broken into five sections. First, we describe the research context including a brief history of EU energy policy. Section 1.2 also reviews existing studies on the topic of EU energy policy and policymaking, Sect. 1.3 introduces and reviews the literature on European interest representation, noting the relative dearth of work on energy and climate policy. Section 1.4 briefly discusses the significance of the burgeoning literature on socio-technical transition and transformation. Section 1.5 summarises and concludes the introduction.

1.2   ENERGY POLICY IN THE EUROPEAN UNION

Human influence on the earth's climate system is occurring, largely as the result of emission of greenhouse gases. The negative consequences of these changes are being felt by societies across the world, and unless emissions are reduced radically, these impacts are very likely to become more severe (IPCC 2014; The Royal Society 2017). The European Union (EU) is the world's third largest emitter of greenhouse gases,[1] 80% of which is associated with energy production, conversion, transportation and use (European Environment Agency 2016).

There can be little doubting that the European Union is a major force in many policy domains from agriculture to international trade, nor that energy has been central to the story of the EU since its earliest days. In the 1950s, creating a common market in coal and steel, the primary basis of military power, under the oversight of a new authority was seen as an important means of ensuring interdependence between the nations of Europe (especially France and Germany) which would make war '*not merely unthinkable, but materially impossible*' (Schuman 1950). But, despite the fact that energy cooperation was at the heart of the Treaty of Paris which established the European Coal and Steel Community (ECSC), the European Commission did not have legal competence to act in the field of energy under the 1957 Treaty of Rome (Lucas 1977). Nevertheless, making use of its competence in other areas, most notably trade harmonisation, a number of energy measures were introduced by the Commission in the 1960s and early 1970s, beginning

---

[1] Behind China and the USA.

with an instrument known as '*the Protocol of Agreement on Energy Problems*' in 1964 (Benson and Russel 2015; Lucas 1977; Weyman-Jones 1986; Cameron 2011, p. 125). Until the late 1980s, however, the EC Commission's Directorate-General responsible for energy was limited to energy research and forecasting (Matlary 1998).

The institutional reform set in train by the Single European Act (SEA) of 1986 and the subsequent drive towards a general European 'internal market' formally brought energy onto the European agenda (Matlary 1998). The inclusion of an energy section in what became the Maastricht Treaty which formally created the European Union in 1992 was proposed but did not make the final text (Duffield and Birchfield 2011). The 1990s saw slow progress in energy integration with a lack of will on the part of the member states to act on proposals by the European Commission preventing a collective response to the energy challenges of the time (Duffield and Birchfield 2011).

The first years of the twenty-first century, however, witnessed a profusion of action on energy. Energy supply concerns, climate change and the drive to create an internal energy market all played a role in raising energy up the EU's agenda. These concerns saw the inclusion of an energy title in 2009s reform to the legal basis of the EU, the Lisbon Treaty.

1.2.1   EU Climate and Energy Policy

Prompted by an emerging understanding of the threat posed by climate change, from the late 1980s the EU began tentative moves towards implementing a Europe-wide policy to limit greenhouse gas (GHG) emissions.

Despite the European Commission's failed attempt to establish a carbon tax in the 1990s and resistance to carbon trading within the 1997 Kyoto Protocol, the idea of a European emissions trading system took hold. In the autumn of 2001, a design for an emissions cap-and-trade system covering installations in the electricity and industrial sectors was proposed and approved. Trading in the EU Emissions Trading System (EU-ETS) began in January 2005 (Wettestad and Jevnaker 2016).

A turning point in the story of energy and the EU was an informal summit at Hampton Court, UK, in 2005 at which, led by UK Prime Minister Tony Blair, leaders of the EU countries agreed that ways towards greater cooperation on energy and climate change should be found (Birchfield 2011; Skjærseth 2013; Eikeland 2012; Mcgowan 2011). The disruption of gas supplies to parts of the community due to a dispute between Russian and Ukrainian gas companies in 2006, as well as the EU's ambitious role as a leader in global climate action also added to a new political commitment to EU energy action. The most visible manifestation of this commitment was the 2020 climate and energy package agreed in 2007. Designed to meet part of the EU's threefold energy goals of clean, secure and competitive energy, the core of the package focused on three targets. There was a commitment to a 20% reduction in emissions (30% in the event of a global deal on climate change), an increase of EU energy consumption from renewable sources to 20% and a 20% improvement in energy efficiency. In 2009, the renewable energy target was implemented in the form of a European Directive, legally binding the member states, to produce a specified proportion of their energy consumption from renewables (European Commission 2009) while energy efficiency legislation has been implemented through a non-binding Directive issued in 2012. The so-called '20/20/20 package' of climate and energy targets has had profound impacts on the energy situation in EU member states. Renewable energy penetration in the energy mix of the EU has increased dramatically with the policies designed to implement the renewable energy target deemed largely responsible (European Commission 2013; Duscha et al. 2016).

While many areas of climate and energy policy remain separate, notably the internal market and the EU's external energy policy, the 20/20/20 package, as it is known, symbolised the partial integration of climate policy and energy policy (Wettestad et al. 2012; Skjærseth 2016; Dupont 2016). The new climate and energy policy area consists primarily of renewable energy policy, energy efficiency policy and the EU-ETS.

In 2009, the UN-administered process for reaching international agreement on climate change saw the 15th 'Conference of the Parties' or COP15, held in Copenhagen, Denmark. At the conference, which was widely hoped would produce a significant deal on global emissions reduction efforts, the EU had high hopes for building on its position as a leader on climate policy (demonstrated by the 20/20/20 package) by helping to ensure a meaningful climate agreement (Dimitrov 2010). Despite the EU's ambition and apparent willingness to act as an international leader on the topic of climate change (Skovgaard 2013, p. 1141), the Copenhagen negotiations did not result in other nations adopting similarly ambitious climate goals. The outcome of the Copenhagen summit is commonly described as unsuccessful or even disastrous for the EU (Skovgaard 2013; Dimitrov 2010).

1.2.2   *Looking Beyond 2020*

Following COP15 in Copenhagen, with the expiry of the 2020 package on the horizon, the terms of a new framework for energy and climate cooperation for the period between 2020 and 2030 emerged. Given the significance of the impact of the 20/20/20 package on energy production and use in EU member states, the targets for 2030 were an important test of Europe's ability to maintain the transformative trajectory established by the 20/20/20 targets. But, despite the similarity in content, negotiating the new package was not a replay of the earlier agreement: The game had changed. This time the European Commission had been granted formal competence in the field of energy by the Lisbon Treaty. Additionally, the Union has grown by three member states and all of the players from national and EU non-governmental and private interests had had seven years to reflect on what works and what does not when negotiating European Climate and Energy policy. At the same time, there were new contextual considerations such as the success of Eurosceptic parties to increase their representation in the European Parliament at the 2014 elections (European Parliament 2014), the ongoing ec, perennial concerns about gas prices and supplies, and questions over Europe's role in global climate negotiations (Groen et al. 2012).

On 23rd October 2014, the heads of state and governments of the 28 member states met at the European Council to decide the content of the 2030 climate and energy policy goals, based on policy proposals from the European Commission. As a result, they agreed a framework within which the EU aims to:

- emit *at least* 40% less greenhouse gas in 2030 than it did in 1990;
- to produce *at least* 27% of the energy consumed in the Union from renewable sources; and
- reduce energy consumption by *at least* 27% compared to a projection derived from a 2007 reference case (European Council 2014).

The conclusions of the European Council were the subject of intense bargaining between member states with the result that various countries extracted concessions. Most strikingly, a group of ten Eastern and

Central European countries including the Bulgaria, Romania and the Visegrad group of Poland, the Czech Republic, Slovakia and Hungary[2] had negotiated what Poland's Prime Minister described as a 'win' (Euractiv 2014). These countries were able to use the threat of a veto of the package to obtain a continued exclusion of parts of their electricity systems from the EU-ETS and access to an energy system modernisation reserve based on ETS revenues (European Council 2014).

1.2.3   The Significance of the 2030 Framework

The scope and structure of the package of targets are notable in two ways. Firstly, the overall level of the targets is lower than many advocates of climate action had hoped. The 'at least' 40% reduction in GHG emissions has been calculated by the European Commission to be the very minimum level commensurate with meeting the target of an 80–95% reduction in emissions (compared to 1990) by 2050, set by the European Council in 2009. The renewable energy target, at 27%, is acknowledged by the European Commission to be more or less what could be expected to be produced by renewables in the absence of any target (European Commission 2014b). The rates of change implied by the 2030 targets are also less ambitious than those implied 20/20/20 package, or the rates of growth achieved to date. The package also rules out individual national targets for energy efficiency and renewable energy with the targets either 'indicative' with no legal weight (in the case of the energy efficiency target) or binding only at a European level (as for the renewable energy target) (European Commission 2014a; European Council 2014).

The move away from national, legally binding renewable energy targets is important for a number of reasons. Firstly, it indicates that ambitious European-level agreement on the goals and means of energy policy has become more difficult since the agreement of the 20/20/20 package in 2007. Secondly, it demonstrates a clear shift towards policy that prefers not to 'pick winners' among energy technologies, placing less emphasis on supporting specific technology groups such as renewables at the European level. By relaxing the requirement for member states to pursue technology-specific policies, the EU pursues a policy, which we

---

[2] Classified in the text as countries with GDP per capita below 60% of the EU average.

argue is inherently more 'technology neutral' (Azar and Sandén 2011) than the 20/20/20 package of policies.

This shift towards a more 'technology-neutral' approach to energy policy is significant because it stands at odds with much of the expanding literature on sustainability transitions and transformations. This literature suggests that a successful energy system transformation is a *'messy, conflictual, and highly disjointed process'* (Meadowcroft 2009, p. 324) and is likely to require policy intervention focussed on specific technologies, rather than simply *'allowing the market'* to choose between a range of low-carbon energy alternatives (Solomon and Krishna 2011). It also shows that the distributional implications of transitions mean that shifting from an 'old' energy system to a 'new' one will always be inherently and unavoidably political (Meadowcroft 2011).

Despite the significance of the targets, the framework continues to evolve. Since 2014, the 2030 targets have been contested, with the level of the energy efficiency target, the renewable energy target, and the overall GHG emissions reduction target the subject of calls for upward revision from, among others, the European Parliament (European Parliament 2017). The nonbinding nature of the renewable energy target also poses a new oversight problem for the European Commission. In the absence of national targets, coordinating the action of member states in pursuit of the goals requires the negotiation and legislation of a new oversight mechanism (European Commission 2017).

1.2.4   Outside Interests and the 2030 Agenda

While the European Council agreed the 2030 climate and energy targets, the choices facing the heads of state in October 2014 were set out in a number of proposals made by the European Commission during that same year with the final package very close in content to that proposed by the Commission. The Commission's proposals, in turn, had been shaped by an intensive period of discussion among a community of policy experts from industry, business and other 'civil society' interests, as well as the staff of the European Commission, which consulted formally on its proposals in 2013 (European Commission 2013).

The nature and impact of this discussion among policy experts in the months and years prior to the European Council in October 2014 is the focus of this book. This early discussion of policies and ideas leading to the selection of topics to be actively considered by policymakers is often described as 'agenda building' or 'agenda-setting' (Cobb and Elder 1972; Jones and Baumgartner 2005).

1.2.5   Studying European Union Climate and Energy Policy

The steady expansion of environmental action taken by the EU over the last few decades has been matched by a parallel growth in literature addressing that activity. Now, as the EU holds some of the most stringent environmental standards in the world, so environmental policy and policymaking is established as a key thread in the literature on European Union politics (Delreux and Happaerts 2016). EU climate policy studies, which emerged as a distinct field of EU environmental politics alongside developments in Brussels, is largely concerned with either the development and functioning of the EU-ETS, the integration of climate policy into other policy fields (Dupont 2016) or the EU's role in the international climate regime (e.g. Groen et al. 2012; Oberthür 2011).

There is no shortage of historical accounts of EU energy policy development (e.g. Matlary 1998; Buchan 2009; Buchan and Keay 2014; Buchan 2013; Kanellakis et al. 2013; Rowlands 2005). Some recent literature looks at EU energy policy through the lenses of European integration, tending to conclude that it is increasingly driven by supranational rather than intergovernmental factors (e.g. Eikeland 2008, 2011; Eising 2002).

There is relatively little work yet that directly addresses the politics of the 2030 targets. The work that has been published emphasises the importance of business interests and powerful pro-business EU officials, as well as mixed political signals from the member states and division within the Commission (Bürgin 2015; Fuchs and Feldhoff 2016). The outcome is generally presented as a negotiated compromise between the various interests (Ydersbond 2016).

1.3   STUDYING EUROPEAN INTEREST GROUPS

This book is primarily interested in the role played by business and civil society interests in shaping the 2030 targets. This section reviews the scholarly field of this kind of 'interest representation', especially in a European context.

The EU has been the subject of academic study since the earliest days of European integration following the Second World War. For many years, scholars focused on understanding the integration phenomenon, with some authors arguing that the European Communities (and later the EU) were an example of the creation of a new kind of supranational polity while others drew attention to the continued domination of the European nation state. See Haas (1958) for an exposition of the former point of view, known as neofunctionalism and Hoffmann (1966) for the basis of the latter, intergovernmentalism. While these intellectual approaches attempted to grapple with the phenomenon of the European integration in order to answer the questions such as 'what is European integration?' and 'how is it possible?' In response to an intellectual impasse in the late 1980s and early 1990s between the two schools, later research tended to take a more comparative approach with a plurality of methodological approaches taken to illuminating particular facets of the EU, such as the policymaking processes by which the EU takes decisions (Rosamond 2000).

The influence of interest groups, defined here as non-governmental actors seeking to influence policy outcomes, and the power of business and civil society in the policy process has been of interest to scholars for at least a century (Woll 2007). Arguably, the topic has been salient for great deal longer than that (e.g. Machiavelli 1532). The 'participatory' rather than exclusively 'representative' role played by non-official actors in a democracy has been discussed for several decades (Pateman 1970; Saurugger 2008). The activity of these interest groups and their relationship to politics and policymaking can be seen as fundamental to questions of democracy and governance (Schmitter 1974, 1977). Crucially, environmental groups became increasingly potent political forces through the 1980s and 1990s (Rawcliffe 1995, 1998) and it has been demonstrated that the activity of interest groups can have a significant impact on the route and rate of energy transitions and transformations (Cherp et al. 2016, 2018).

Classical models of 'interest representation' or 'interest intermediation' can be grouped into two basic theoretical categories: corporatism and pluralism (Schmitter 1974), both alternatives to statism (Kohler-Koch and Eising 2003). Corporatism assumes the granting of monopolies of representation to a limited number of private interests by the state, while pluralism assumes that interests are negotiated among an unspecified number of organisations, with no monopoly granted or licenced by the state (see Dryzek and Dunleavy 2009 for a good overview of the concept; as well as Dahl 1978). Later 'neo-pluralist' thinking about the nature of pluralism holds that, rather than treating all interests symmetrically, the existence of a market system leads to a situation in which business or capital, represented by the interests of corporations, often wields more power over policy than other interests (Lindblom 1982; McFarland 2007).

While we start this study from an agnostic position about the relative power of different classes of actor, it is clear that the threshold to involvement in the EU policymaking is rather high and policy process is more accessible to some people than to others. The relationship between the institutions of the EU and external interests is sometimes described as a form of '*élite pluralism*' in which access is not restricted or managed by the institutions. However, it is '*a system where access is generally restricted to a few policy players* [or élites], *for whom membership is competitive and strategically advisable, but not compulsory or enforceable as* [it is in] *the corporate model*' (Coen 1997, p. 98).

A great deal of effort has been expended trying to understand the *power* of interest groups and how it is wielded (e.g. Gray and Lowery 1996; Lowery and Gray 1998, 2004, 2005; Balanyá et al. 2003; Dür and De Bièvre 2007; Mazey and Richardson 2001). Baumgartner et al. (2009) provide an especially comprehensive survey of interest group access, strategies and success factors in the USA.

Nevertheless, power has been shown to be expressed in more nuanced ways than a simple expression of 'winners and losers' and is difficult or impossible to observe directly. Instead, a fruitful research agenda might focus on systematically observing some of the elements that are known to contribute to power such as access to policymakers and actors' resources (Woll 2007; Klüver 2011).

It can be observed that, due to the fact that access to policy engagement and influence tends to be restricted to formal and legislative bodies during the later stages of policy formulation, legislation and implementation, outside interests tend to have the greatest impact during the earliest phase of a policy's life (Birkland 2010; Princen 2007; Baumgartner et al. 2009). This period is often described as the 'agenda-setting' phase and it is this phase of the policy process that we take as the focus of this book.

The institutions of the EU, compared to the population which it serves, are relatively underfunded and often require outside input in many policy areas, leading to intense relationships between policymakers and interest groups, the interests of which are taken as a proxy for the interests of the wider population (Majone 1994; Richardson and Coen

2009). Interest groups have sought to influence European-level policy-making since the birth and development of the three original European Communities in the 1950s (Haas 1958). The inception of the single market in the 1980s also prompted growth in interest group activity relative to European policymaking (Greenwood and Aspinwall 1998).

EU lobbying activity itself has been accompanied by similar growth in studies of interest group activity, both sectoral, empirical overviews and theoretical contributions with debates about the nature and implications of this type of engagement still contentious (Richardson and Coen 2009; Greenwood 2011; Bouwen 2002; Coen 2007; Adelle and Anderson 2013; Curtin 2003; Mahoney 2007). There is also a thriving business in producing 'how-to' guides for EU lobbyists (e.g. de Cock 2010; Geiger 2012; van Schendelen 2013). Wide variations are observed between policy areas, institutional venues, organisation type, stage of the policy process, type of actor and other factors, making systematic understanding of the phenomenon of interest representation challenging (Greenwood 2011; Richardson and Coen 2009; Long and Loerinczi 2009; Warleigh 2006; Baumgartner et al. 2009).

Contrary to the experience of the Federal US setting where much theorising about the role of interest groups in agenda-setting has taken place and, indeed, that in EU member states, agenda-setting in the EU tends to be more of an 'insider' affair, excluding the public mobilisation by lobbyists characteristic of national policymaking (Princen 2007, 2018). This is likely a result of the absence of an 'integrated EU public sphere' (Herweg 2016). For this reason, we focus primarily on the insider lobbying activity of 'European interest groups', that is, those civil society organisations which are based and carry out their work in Brussels, excluding the somewhat larger group of primarily national organisations which may use public-facing strategies for influence.

1.4 TECHNOLOGY, TRANSITIONS AND TRANSFORMATIONS

To provide an account of a policymaking process, it is important that we consider the fundamental objectives of the policy. In the case of the 2030 targets, the ultimate aim is to contribute to Europe's 'transition' to, or transformation into, a sustainable, low-carbon economy (European Commission 2014a). This section provides an overview of the key concepts contained in the literature about sustainability transitions and transformations. It also introduces and describes the idea of technology neutrality, which later chapters will show to be an important factor in the policy debate about the 2030 targets.

Achieving the EU's stated goal of near complete decarbonisation by 2050 will require a radical change in the way that energy is produced and consumed (European Council 2009). While some steps can be made through incremental means such as the improved operation of existing technology, the scale of change required to reduce emissions as far and as fast as has been indicated to be necessary demands 'system innovation' (Elzen et al. 2004). In other words, the *relatively stable configurations of institutions, techniques and artefacts* or 'socio-technical system' (STS) that encompasses the various elements of the energy system will need to *'transition'* from its current state to some sustainable future state (Geels 2002; Kern and Howlett 2009). The 2030 targets' stated goal is to contribute to the transition of the EU's energy system (European Commission 2014a).

To some, the concept of a 'singular' or 'scalar' *transition* from one state to another through time is reductive and unable to fully reflect the implications of uncertainty, power and diversity inherent in the move to environmental sustainability. Instead the concept of a plural, vector or open *transformation* is invoked in which progress is *'best represented, not as a single-track 'race', but as palimpsests of branching counterfactual paths'* (Stirling 2010, 2011, p. 83; 2014). For the purposes of this book, however, the distinction between transition and transformation is somewhat less relevant with the critical point being that major change is required and that the specific policy action, including at the EU level, can and is playing a part in enabling or restraining that change. A growing literature on the topic of socio-technical transitions and transformations underpins several assertions about their nature relevant to this study:

- First, analysis of historical transitions demonstrates they do not happen by accident and rarely, if ever, happen quickly (Solomon and Krishna 2011);
- Second, an important 'steering' role for public authorities is ascribed. This role is especially important in overcoming 'lock-in' or returns to scale of dominant unsustainable technologies or practices (Verbong and Geels 2007; Kern and Howlett 2009); and
- Third, distributional impacts and implications for transitions mean that the successful navigation of path to a sustainable energy system will be inherently political. There may be losers as well as winners.

Factors such as interests, institutions and ideas matter in the same way as policy does in determining transitional or transformational outcomes (Meadowcroft 2009, 2011; Stirling 2014).

These insights tell us firstly that the swing in EU climate and energy policy towards a less managed, market-driven transition pathway may have significant implications for the effectiveness of the transition European politicians claim to be pursuing with a technology neutral approach to policy less likely to bring about the scale or rate of change required (Azar and Sandén 2011). The impact of the 20/20/20 targets from 2007 on renewable energy growth shows how potent the EU-level steering role can be (European Environment Agency 2014).

1.5 Conclusion

Having provided an account of the policy and research context, this chapter surveyed several areas of pertinent background academic literature. It has showed that while EU climate and energy policy is a thriving research area, which has grown as the policy has developed, it has been overlooked by studies of the politics of the policy process and little work has been carried out which explicitly examines the process by which the 2030 targets were selected.

A plurality of analytical approaches can be used to illuminate issues of EU political life and policymaking. Among these is the study of European interest representation. This field is highly fragmented, with a range of approaches adopted and a large number of case studies carried out across policy areas and issues. Despite this growing body of literature, the role of interest groups in climate and energy policy has often been overlooked and remains relatively under-researched.

The study of socio-technical transitions is a varied and rapidly growing field. The literature shows that not only are the 2030 targets effectively a decision to divert from the sustainable path established by the 20/20/20 framework but that the nature of the transition Europe's energy system must undergo means that political factors are inevitably important.

This review has identified that gaps in the scholarship relating EU studies, socio-technical transitions and interest representation offer a fertile area for research. This book addresses those lacunae and provides a novel approach in so far as it combines insights from all three bodies of literature.

REFERENCES

Adelle, C., & Anderson, J. (2013). Lobby Groups. In A. Jordan & C. Adelle (Eds.), *Environmental Policy in the EU: Actors, Institutions and Processes*. London and New York: Routledge.

Azar, C., & Sandén, B. A. (2011). The Elusive Quest for Technology-Neutral Policies. *Environmental Innovation and Societal Transitions, 1*(1), 135–139.

Balanyá, B., et al. (2003). *Europe Inc: Regional and Global Restructuring and the Rise of Corporate Power* (2nd ed.). London: Pluto Press.

Baumgartner, F. R., et al. (2009). *Lobbying and Policy Change: Who Wins, Who Loses, and Why*. Chicago, IL: University of Chicago Press.

Benson, D., & Russel, D. (2015, January). Patterns of EU Energy Policy Outputs: Incrementalism or Punctuated Equilibrium? *West European Politics, 38*(1), 37–41.

Birchfield, V. L. (2011). The Role of EU Institutions in Energy Policy Formation. In V. L. Birchfield & J. Duffield (Eds.), *Toward a Common European Union Energy Policy* (pp. 235–262). London and New York: Palgrave Macmillan.

Birkland, T. A. (2010). *An Introduction to the Policy Process: Theories, Concepts and Models of Public Policy Making*. Oxford: Routledge.

Bouwen, P. (2002). Corporate Lobbying in the European Union: The Logic of Access. *Journal of European Public Policy, 9*(3), 365–390.

Buchan, D. (2009). *Energy and Climate Change: Europe at the Crossroads*. Oxford: Oxford University Press.

Buchan, D. (2013). *Why Europe's Energy and Climate Policies are Coming Apart*. Available at: https://www.oxfordenergy.org/wpcms/wp-content/uploads/2013/07/SP-28.pdf. Accessed 20 Mar 2014.

Buchan, D., & Keay, M. (2014). *The EU's New Energy and Climate Goals for 2030: Under-Ambitious and Over-Bearing?* Available at: https://www.oxfordenergy.org/wpcms/wp-content/uploads/2014/01/The-EUs-new-energy-and-climate-goals-for-2030.pdf. Accessed 20 Mar 2014.

Bürgin, A. (2015). National Binding Renewable Energy Targets for 2020, but Not for 2030 Anymore: Why the European Commission Developed from a Supporter to a Brakeman. *Journal of European Public Policy, 22*(5), 690–707.

Cameron, P. D. (2011). The EU and Energy Security: A Critical Review of the Legal Issues. In A. Antoniadis, R. Schütze, & E. Spaventa (Eds.), *The European Union and Global Emergencies: A Law and Policy Analysis* (pp. 125–166). Portland: Bloomsbury.

Cherp, A., et al. (2016, May). Comparing Electricity Transitions: A Historical Analysis of Nuclear, Wind and Solar Power in Germany and Japan. *Energy Policy, 101*, 612–628.

Cherp, A., et al. (2018, September 2017). Integrating Techno-economic, Socio-technical and Political Perspectives on National Energy Transitions: A Meta-theoretical Framework. *Energy Research and Social Science, 37,* 175–190.

Cobb, R., & Elder, C. D. (1972). *Participation in American Politics: The Dynamics of Agenda Building.* Boston: Allyn and Bacon.

de Cock, C. (2010). *iLobby.eu: Survival Guide to EU Lobbying.* Delft: Eburon.

Coen, D. (1997). The Evolution of the Large Firm as a Political Actor in the European Union. *Journal of European Public Policy, 4*(1), 91–108.

Coen, D. (2007). Empirical and Theoretical Studies in EU Lobbying. *Journal of European Public Policy, 14*(3), 333–345.

Curtin, D. (2003). Private Interest Representation or Civil Society Deliberation? A Contemporary Dilemma for European Union Governance. *Social and Legal Studies, 12,* 55–75.

Dahl, R. A. (1978). Pluralism Revisited. *Comparative Politics, 10*(2), 191–203.

Delreux, T., & Happaerts, S. (2016). *Environmental Policy and Politics in the European Union.* London: Palgrave Macmillan.

Dimitrov, R. S. (2010). Inside UN Climate Change Negotiations: The Copenhagen Conference. *Review of Policy Research, 27*(6), 795–821.

Dryzek, J., & Dunleavy, P. (2009). *Theories of the Democratic State.* Basingstoke: Palgrave Macmillan.

Duffield, J., & Birchfield, V. L. (2011). Toward a Common European Union Energy Policy. In J. S. Duffield & V. L. Birchfield (Eds.), *Problems, Progress, and Prospects* (235–262). Basingstoke: Palgrave Macmillan.

Dupont, C. (2016). *Climate Policy Integration into EU Energy Policy: Progress and Prospects.* Oxford: Routledge.

Dür, A., & De Bièvre, D. (2007). The Question of Interest Group Influence. *Journal of Public Policy, 27*(1), 1–12.

Duscha, V., Held, A., & del Rio, P. (2016). An Economic Analysis of the Interactions Between Renewable Support and Other Climate and Energy Policies. *Energy & Environment, 28*(1–2), 1–23.

Eikeland, P. O. (2008). *EU Internal Energy Market Policy: New Dynamics in the Brussels Policy Game?* Available at: https://www.fni.no/getfile.php/132068/Filer/Publikasjoner/FNI-R1408.pdf. Accessed 13 Feb 2014.

Eikeland, P. O. (2011). The Third Internal Energy Market Package: New Power Relations Among Member States, EU Institutions and Non-State Actors? *Journal of Common Market Studies, 49*(2), 243–263.

Eikeland, P. O. (2012). *EU Energy Policy Integration—Stakeholders, Institutions and Issue-Linkages.* Available at: https://www.fni.no/getfile.php/132050/Filer/Publikasjoner/FNI-R1312.pdf. Accessed 13 Feb 2014.

Eising, R. (2002). Policy Learning in Embedded Negotiations: Explaining EU Electricity Liberalization Policy Learning in Embedded Negotiations: Explaining EU Electricity Liberalization. *International Organization, 56*(1), 85–120.

Elzen, B., Geels, F. W., & Green, K. (2004). *System Innovation and the Transition to Sustainability*. Cheltenham: Edward Elgar.

Euractiv. (2014). Poland says it "won" at the EU summit. Available at: http://www.euractiv.com/sections/energy/poland-says-it-won-eu-summit-309494. Accessed 24 Nov 2014.

European Commission. (2009). *Directive 2009/72/EC Concerning Common Rules for the Internal Market in Electricity and Repealing Directive 2003/54/EC*. Available at: http://eur-lex.europa.eu/legal-content/EN/TXT/PDF/?uri=CELEX:32003L0054&from=EN. Accessed 13 Feb 2014.

European Commission. (2013). *Green Paper: A 2030 Framework for Climate and Energy Policies*. Available at: http://ec.europa.eu/transparency/regdoc/rep/1/2013/EN/1-2013-169-EN-F1-1.pdf. Accessed 10 Feb 2014.

European Commission. (2014a). *Impact Assessment—A Policy Framework for Climate and Energy in the Period from 2020 to 2030*. Available at: http://eur-lex.europa.eu/legal-content/EN/TXT/PDF/?uri=CELEX:52014DC0015&from=EN. Accessed 4 Feb 2014.

European Commission. (2014b). *Minutes of the 2072nd Meeting of the Commission Held in Brussels (Berlaymont) on Wednesday 22 January*. Available at: http://ec.europa.eu/transparency/regdoc/rep/10061/2014/EN/10061-2014-2072-EN-F1-1.Pdf. Accessed 20 Apr 2016.

European Commission. (2017). *Proposed Regulation on the Governance of the Energy Union*. Available at: http://eur-lex.europa.eu/resource.html?uri=cellar:f9f04518-b7dc-11e6-9e3c-01aa75ed71a1.0001.02/DOC_1&format=PDF. Accessed 22 Feb 2018.

European Council. (2009). *29/30 October 2009: Conclusions*. Available at: http://www.consilium.europa.eu/uedocs/cms_data/docs/pressdata/en/ec/110889.pdf. Accessed 31 Mar 2016.

European Council. (2014). *European Council (23 and 24 October 2014) Conclusions on 2030 Climate and Energy Policy Framework*. Available at: http://www.consilium.europa.eu/uedocs/cms_data/docs/pressdata/en/ec/145397.pdf. Accessed 24 Oct 2014.

European Environment Agency. (2014). *Trends and Projections in Europe 2014: Tracking Progress Towards Europe's Climate and Energy Targets for 2020*. Available at: https://www.eea.europa.eu/publications/trends-and-projections-in-europe-2014/at_download/file. Accessed 28 Oct 2014.

European Environment Agency. (2016). *Annual European Union Greenhouse Gas Inventory 1990–2012 and Inventory Report 2014*. Available at: http://www.eea.europa.eu/publications/european-union-greenhouse-gas-inventory-2014. Accessed 22 Feb 2018.

European Parliament. (2014). *European Elections 2014*. Available at: http://www.europarl.europa.eu/us/en/elections_2014.html;jsessionid=ABD017EA7C44EE440340530246C59FAA. Accessed 3 June 2014.

European Parliament. (2017). *Minutes: Committee on Industry, Research and Energy, Meeting of 27 November 2017, 15.00–18.30, and 28 November 2017, 9.00–12.30 and 14.30–18.30.* Available at: http://www.europarl.europa.eu/sides/getDoc.do?type=COMPARL&reference=PE-615.215&format=PDF&language=EN&secondRef=01. Accessed 22 Feb 2018.

Fuchs, D., & Feldhoff, B. (2016). Passing the Scepter, Not the Buck: Long Arms in EU Climate Politics. *Journal of Sustainable Development, 9*(6), 58.

Geels, F. W. (2002). Technological Transitions as Evolutionary Reconfiguration Processes: A Multi-Level Perspective and a Case-Study. *Research Policy, 31*(8–9), 1257–1274.

Geiger, A. (2012). *EU Lobbying Handbook* (2nd ed.). CreateSpace Independent Publishing Platform.

Gray, V., & Lowery, D. (1996). A Niche Theory of Interest Representation. *The Journal of Politics, 58*(1), 91–111.

Greenwood, J. (2011). *Interest Representation in the European Union.* Basingstoke: Palgrave Macmillan.

Greenwood, J., & Aspinwall, M. D. (1998). *Collective Action in the European Union: Interests and the New Politics of Associability.* London and New York: Routledge.

Groen, L., Niemann, A., & Oberthür, S. (2012). The EU as a Global Leader? The *Copenhagen and Cancun UN Climate Change Negotiations, 8*(2), 173–191.

Haas, E. B. (1958). *The Uniting of Europe: Political, Social and Economic Forces, 1950–1957.* Stanford, CA: Stanford University Press (Reprint, Notre Dame, IN: University of Notre Dame Press, 2003).

Herweg, N. (2016). Explaining European Agenda-Setting Using the Multiple Streams Framework: The Case of European Natural Gas Regulation. *Policy Sciences, 49*(1), 13–33.

Hoffmann, S. (1966). Obstinate or Obsolete? The Fate of the Nation-State and The Case of Western Europe. *Daedalus, 95*(3), 862–915.

IPCC. (2014). *Climate Change 2014 Synthesis Report Summary Chapter for Policymakers.* Available at: https://www.ipcc.ch/pdf/assessment-report/ar5/syr/AR5_SYR_FINAL_SPM.pdf. Accessed 15 Aug 2016.

Jones, B. D., & Baumgartner, F. R. (2005). *The Politics of Attention: How Government Prioritizes Problems.* Chicago: University of Chicago Press.

Kanellakis, M., Martinopoulos, G., & Zachariadis, T. (2013). European Energy Policy—A Review. *Energy Policy, 62*, 1020–1030.

Kern, F., & Howlett, M. (2009). Implementing Transition Management as Policy Reforms: A Case Study of the Dutch Energy Sector. *Policy Sciences, 42*(4), 391–408.

Klüver, H. (2011). *Lobbying in Coalitions: Interest Group Influence on European Union Policy-Making.* Available at: https://www.nuffield.ox.ac.uk/politics/papers/2011/HeikeKluever_workingpaper_2011_04.pdf. Accessed 15 Aug 2016.

Kohler-Koch, B., & Eising, R. (2003). *Transformation of Governance in the European Union.* London and New York: Routledge.

Lindblom, C. E. (1982). The Market as Prison. *The Journal of Politics, 44*(2), 324–336.

Long, T., & Loerinczi, L. (2009). NGOs as Gatekeepers: A Green Vision. In D. Coen & J. Richardson (Eds.), *Lobbying the European Union: Institutions, Actors, and Issues* (pp. 169–188). Oxford: Oxford University Press.

Lowery, D., & Gray, V. (1998). The Dominance of Institutions in Interest Representation: A Test of Seven Explanations. *American Journal of Political Science, 42*(1), 231–255.

Lowery, D., & Gray, V. (2004). A Neopluralist Perspective on Research on Organized Interests. *Political Research Quarterly, 57*(1), 164–175.

Lowery, D., & Gray, V. (2005). Sisyphus Meets the Borg. *Journal of Theoretical Politics, 17*(1), 41–74.

Lucas, N. J. D. (1977). *Energy and the European Communities.* London: Europa Publications.

Machiavelli, N. (1532). *The Prince.* London: Penguin (Reprint, Chicago: University of Chicago Press, 2010).

Mahoney, C. (2007). Lobbying Success in the United States and the European Union. *Journal of Public Policy, 27*(1), 35–56.

Majone, G. (1994). The Rise of the Regulatory State in Europe. *West European Politics, 17*(3), 77–101.

Matlary, J. H. (1998). *Energy Policy in the European Union.* Basingstoke: Palgrave Macmillan.

Mazey, S., & Richardson, J. (2001). Interest Groups and EU Policy-Making: Organisational Logic and Venue Shopping. In J. Richardson (Ed.), *European Union: Power and Policy-Making* (pp. 247–268). New York: Routledge.

McFarland, A. S. (2007). Neopluralism. *Annual Review of Political Science, 10*(1), 45–66.

McGowan, F. (2011). The UK and EU Energy Policy: From Awkward Partner to Active Protagonist? In V. L. Birchfield & J. Duffield (Eds.), *Toward a Common European Union Energy Policy* (pp. 187–213). New York: Palgrave Macmillan.

Meadowcroft, J. (2009). What About the Politics? Sustainable Development, Transition Management, and Long Term Energy Transitions. *Policy Sciences, 42*(4), 323–340.

Meadowcroft, J. (2011). Engaging with the Politics of Sustainability Transitions. *Environmental Innovation and Societal Transitions, 1*(1), 70–75.

Oberthür, S. (2011). The European Union's Performance in the International Climate Change Regime. *Journal of European Integration, 33,* 667–682.

Pateman, C. (1970). *Participation and Democratic Theory.* Cambridge: Cambridge University Press.

Princen, S. (2007). Agenda-setting in the European Union: A Theoretical Exploration and Agenda for Research. *Journal of European Public Policy*, *14*(1), 21–38.

Princen, S. (2018). Agenda-Setting and Framing in Europe. *The Palgrave Handbook of Public Administration and Management in Europe* (pp. 535–551). London: Palgrave Macmillan.

Rawcliffe, P. (1995). Making Inroads: Transport Policy and the British Environmental Movement. *Environment*, *37*(3), 16–36.

Rawcliffe, P. (1998). *Environmental Pressure Groups in Transition*. Manchester: Manchester University Press.

Richardson, J., & Coen, D. (2009). *Lobbying the European Union: Institutions, Actors, and Issues*. Oxford: Oxford University Press.

Rosamond, B. (2000). *Theories of European Integration*. Basingstoke and New York: Palgrave Macmillan.

Rowlands, I. H. (2005). The European Directive on Renewable Electricity: Conflicts and Compromises. *Energy Policy*, *33*(8), 965–974.

Saurugger, S. (2008). Interest Groups and Democracy in the European Union. *West European Politics*, *31*(6), 1274–1291.

van Schendelen, M. P. C. M. (2013). *The Art of Lobbying the EU: More Machiavelli in Brussels*. Amsterdam: Amsterdam University Press.

Schmitter, P. C. (1974). Still the Century of Corporatism? *The Review of Politics*, *36*(1), 85–131.

Schmitter, P. C. (1977). Modes of Interest Intermediation and Models of Societal Change in Western Europe. *Comparative Political Studies*, *10*(1), 7–38.

Schuman, R. (1950). *Schuman Declaration*. Available at: https://europa.eu/european-union/about-eu/symbols/europe-day/schuman-declaration_en. Accessed 31 Oct 2016.

Skjærseth, J. B. (2013). *Unpacking the EU Climate and Energy Package: Causes, Content and Consequences*. Available at: https://www.fni.no/getfile.php/131681/Filer/Publikasjoner/FNI-R0213.pdf. Accessed 16 Jan 2015.

Skjærseth, J. B. (2016). Linking EU Climate and Energy Policies: Policy-Making, Implementation and Reform. *International Environmental Agreements: Politics, Law and Economics*, *16*(4), 509–523.

Skovgaard, J. (2013). The Limits of Entrapment: The Negotiations on EU Reduction Targets, 2007–11. *Journal of Common Market Studies*, *51*(6), 1141–1157.

Solomon, B. D., & Krishna, K. (2011). The Coming Sustainable Energy Transition: History, Strategies, and Outlook. *Energy Policy*, *39*(11), 7422–7431.

Stirling, A. (2010). Keep it Complex. *Nature*, *468*(7327), 1029–1031.

Stirling, A. (2011). Pluralising Progress: From Integrative Transitions to Transformative Diversity. *Environmental Innovation and Societal Transitions,* *1*(1), 82–88.

Stirling, A. (2014). Transforming Power: Social Science and the Politics of Energy Choices. *Energy Research and Social Science, 1,* 83–95.

The Royal Society. (2017). *Climate Updates: What Have We Learnt Since the IPCC 5th Assessment Report?.* London: The Royal Society.

Verbong, G., & Geels, F. W. (2007). The Ongoing Energy Transition: Lessons from a Socio-technical, Multi-level Analysis of the Dutch Electricity System (1960–2004). *Energy Policy, 35*(2), 1025–1037.

Warleigh, A. (2006). Making Citizens from the Market? NGOs and the Representation of Interests. In R. Bellamy, D. Castiglione, & J. Shaw (Eds.), *Making European Citizens: Civic Inclusion in a Transnational Context* (pp. 118–132). London: Palgrave Macmillan.

Wettestad, J., Eikeland, P. O., & Nilsson, M. (2012). EU Climate and Energy Policy: A Hesitant Supranational Turn? *Global Environmental Politics, 12*(2), 67–86.

Wettestad, J., & Jevnaker, T. (2016). *Rescuing EU Emissions Trading: The Climate Policy Flagship.* London: Palgrave Macmillan.

Weyman-Jones, T. G. (1986). *Energy in Europe: Issues and Policies.* London and New York: Methuen.

Woll, C. (2007). Leading the Dance? Power and Political Resources of Business Lobbyists. *Journal of Public Policy, 27*(1), 57.

Ydersbond, I. M. (2016). *Where Is Power Really Situated in the EU?* Oslo: Fridtjof Nansen Institute.

CHAPTER 2

Analytical Framework

Abstract In this chapter, key concepts and terms are defined and an analytical framework based on the multiple streams approach (MSA) developed by John Kingdon in the 1980s is introduced and described. MSA is argued to be an especially suitable conceptual approach to the analysis of climate and energy policy interest group activity and that it is especially well suited to application in an EU context.

Keywords Multiple streams approach · European Union policymaking Policy process · Policy windows · Policy entrepreneurship

2.1 INTRODUCTION: SELECTING A FRAMEWORK

The purpose of this book is analyse the agenda-setting processes that preceded formal decisions relating to the 2030 climate and energy framework. In this chapter, we develop the analytical framework used to guide the study and structure this book. First, we set out the key concepts and definitions used throughout the book. Section 2.2 then presents and depicts the analytical framework or 'lens' through which the subject, the debate about the 2030 targets in Brussels, is scrutinised. We contend that using the multiple streams approach (MSA) for understanding the policy process will shed new light on the research topic and extend existing scholarship in this area. Section 2.3 concludes the chapter.

© The Author(s) 2018   21
O. Fitch-Roy and J. Fairbrass, *Negotiating the EU's 2030
Climate and Energy Framework*, Progressive Energy Policy,
https://doi.org/10.1007/978-3-319-90948-6_2

As with many processes of policymaking, perhaps especially in the EU, agenda-setting is extremely complex (Tosun et al. 2015). To simplify and understand complex phenomenon, typically as researchers, we begin with a set of presuppositions about the processes involved. Such presuppositions inform our view of the world whether we explicitly acknowledge and define them or not. So, we face a choice: To proceed with a 'common sense' approach reliant on the categories and assumptions that have arisen from our own experience, which may be riven by *'internal inconsistencies, ambiguities, erroneous assumptions, and invalid propositions'* or to proceed on the basis of *'clear and logically interrelated propositions'* (Sabatier 2007, p. 5). This thesis adopts the second approach. This chapter presents the choices made to establish an explicit framework that at once focuses the analysis on the most important elements of the agenda-setting process and defines the categories into which the findings can usefully be grouped.

There are a great many frameworks for analysis of the policy process from which to choose. Weible and Sabatier (2017) outline a selection of seven distinct, qualified approaches, specifically chosen because they were then being developed, discussed and applied. Each one is supported by its own theoretical and empirical literature. At the same time, institutional approaches, especially the Historical and Discursive variants, have been used to analyse various energy policy processes (Palmer 2010; Lorenzoni and Benson 2014). However, the selection of approaches presented by Weible and Sabatier are vetted for clarity, requiring that proponents show them to be *'clear enough to be proven wrong'* and their popularity means that there is a large body of literature on which to draw.

Of these, two in particular have features that may lend themselves to our chosen topic: The MSA and the advocacy coalition framework (ACF). Both approaches seek to explain outcomes in the public policymaking process in which technical information, the personal beliefs of participants and large-scale political factors are important factors in explaining outcomes (Sabatier and Jenkins-Smith 1999; Kingdon 2010). They also emphasise the role that communities of experts (or, in the language of ACF, policy subsystems), including civil society actors, play in negotiating outcomes (Howlett et al. 2016). However, while ACF is an attempt to create a model of the entire policy process, MSA was developed with agenda-setting in mind (Kingdon 2010; Zahariadis 2014). Since our research focuses tightly on the agenda-setting process, MSA has been selected as the analytical framework used here.

2.2   STRUCTURE OF MSA

The MSA is the application of the concept of the 'garbage can' model of organisation choice proposed by Cohen et al. (1972) to the field of political science. Originally developed by John Kingdon in 1984, specifically to explain the agenda-setting process, MSA has inspired tens of thousands of publications, empirical and theoretical, covering dozens of countries and policy domains (Zahariadis 2014; Cairney and Jones 2016; Jones et al. 2016). MSA is still very much an active topic in policy theory (Herweg 2015, 2016; Winkel and Leipold 2016; Herweg et al. 2015; Weible and Schlager 2016; Cairney and Jones 2016; Herweg and Zahariadis 2018).

The enduring appeal of MSA is its ability to see problems, potential policy solutions and the political context developing largely independently, only to be brought together by policy entrepreneurs, resourceful and talented actors who can 'do more with less', during brief critical junctures known as 'policy windows' (Kingdon 2010; Zahariadis 2014; Mintrom and Norman 2009). Crucially, the conceptual separation of problems and policies allows for solutions to pre-date problems, with advocates often seeking to strategically present the same *solution* to a variety of salient *problems* over time (Kingdon 2010).

The MSA has five elements that interact to produce policy outcomes: The three streams of 'policy', 'problem' and 'politics', which are assumed to flow through the decision-making system independently of one another. 'Policy windows' are opportunities that occur when the conditions in the streams are ripe for skilled 'policy entrepreneurs' to couple the streams and affect change in the agenda. Figure 2.1 shows the structure of MSA as applied here in schematic including some augmentation of the role of the policy entrepreneur, discussed in the following pages. The following subsections examine each of the structural elements in turn (Fig. 2.1).

2.2.1   *Problem Stream*

The problem stream consists of the conditions that policymakers, on behalf of citizens, would like to address. These may include issues like climate change, energy security, economic crisis or the cost of energy

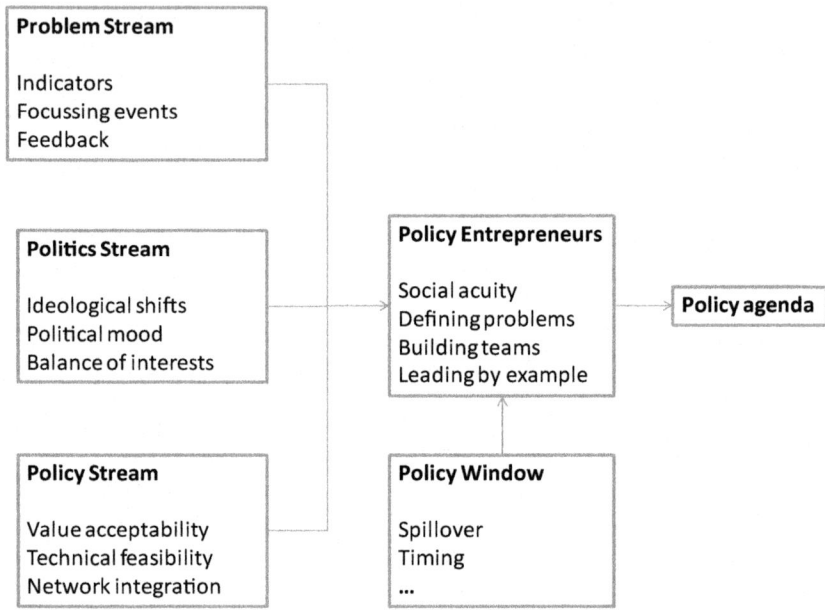

Fig. 2.1 The structure of the multiple streams approach (MSA) (authors' adaptation from Zahariadis 2014; Mintrom and Norman 2009)

for households and businesses. It is important to recognise that 'conditions' do not become 'problems' automatically and that problem definition is an important means by which actors seek to influence the agenda (Rochefort and Cobb 1994; Princen 2018). Kingdon (2010) proposes that problems are framed and receive attention in three separate ways— *as indicators, via feedback and focusing events.*

2.2.2   The Policy Stream

The policy stream is a 'primaeval soup' of ideas vying for acceptance within the relevant policy community of specialists involved in a particular policy area, in this case EU climate and energy policy (Zahariadis 2014; Richardson 2005). Kingdon (2010) and Zahariadis (2014) both describe the policy community as networks of officials, academics, think tanks and other researchers. However, as demonstrated in later chapters of this book and in line with Rozbicka and Spohr's (2016) article looking at EU chemical's policy, interest groups, such as campaign organisations, NGOs and trade associations are also important members of the policy community and are able, through political manipulation of information and persuasion, able to wield considerable power (Zahariadis 2014). It is this community, organised as a network, which allows ideas and issues to be handled in parallel rather than series, to some extent mitigating the effect of time scarcity (Zahariadis 2014). The term 'policy subsystem' is sometimes used to capture the concept (Baumgartner and Jones 1991; McCool 1998; Sabatier and Jenkins-Smith 1999; Sabatier 1998; Börzel 1998).

The rate at which a policy idea rises or falls on the agenda within a community is a function of the properties of the particular policy network. More integrated networks, which are smaller and more consensual with restricted access, tend to see ideas emerge rather faster than in larger, more competitive and less restricted networks. Kingdon also stresses the importance of a process he calls 'softening up' (Kingdon 2010, p. 127). By exposing both policy communities and the wider public to new policy ideas, advocates 'prepare the ground' for an opportunity to link problem and policy.

2.2.3   The Politics Stream

The politics stream is the wider political environment in which the policy process is embedded. It includes the ideological tendencies of EU member state governments, the preferences of the political leaders of the European Commission and inter-institutional issues such as the balance of power in and between the Council and Parliament. Kingdon, as well as later contributors to the MSA, make the case that interest group activity is part of this political stream (Zahariadis 2014; Ackrill et al. 2013). The work of Rozbicka and Spohr (2016), however, makes a case for a larger role for interest group activity across all three streams. Interest groups may be engaged in framing in the problem stream, developing and negotiating proposals in the policy stream and skilfully deploying resources at the right time to make the most out of windows of opportunity in the politics stream.

Whether an issue receives attention or an idea will gain traction depends, to some extent, on the political 'mood' (Kingdon 2010; Zahariadis 2014; Ackrill and Kay 2011; Ackrill et al. 2013). For example, at a time when the European *zeitgeist* is one of austerity and fiscal prudence, policy ideas that are presented as more economically 'efficient' or lower cost are more likely to receive approval in political circles (Zahariadis 2007; Kingdon 2010).

In a national or federal context, the MSA tends to include three main elements within the politics stream: The national mood, pressure group campaigns and administrative or legislative turnover. The unique political environment in the EU means that some modifications are needed to translate the framework to an EU context. While various approaches have been taken by other authors (Herweg and Zahariadis 2018), we follow Zahariadis (2008, p. 518) suggestion that '*the balance of Council member national and partisan affiliation, the ideological balance of parties in Parliament, and the European mood*' make up the EU politics stream.

2.2.4   Policy Windows

Policies are selected and applied to problems during moments Kingdon describes as 'policy windows'. These windows of opportunity are open for a finite duration and see advocates of particular policies push their proposals or bring attention to a specific problem. Windows may open to due unpredictable external events such as large-scale electrical power outages affecting the politics of the situation or may open on a regular basis due to entirely predictable legislative or other timetables. The bringing together of policy and problems streams—known as coupling—occurs only during a policy window. In MSA, it is assumed that policy actors themselves do not control when or how wide the window opens. The hypothesis is that policy actors are somewhat like 'surfers' lying in wait for waves (Boscarino 2009; Kingdon 2010).

Kingdon (2010) borrows the term 'spillover' from Ernst Haas' (1958) neofunctionalist approach to European integration in describing the effect that action in one policy arena may have on another. Kingdon suggests that spillover may occur when a principle is established or a precedent set which alters the terms under which a policy topic is discussed in the future. Kingdon also suggests that success by a policy entrepreneur in one area may affect what happens in other, adjacent or overlapping policy area. This could be because a particular policy idea, once shown to be successful in one field then becomes attractive to entrepreneurs seeking success in another. It could also be the case that '…*the coalition that was built and nurtured to establish the new policy can be transferred to other fights*' (Kingdon 2010, p. 192).

The spillover mechanism, although loosely defined in Kingdon's original 'precedent-setting' formulation, has received more recent attention. Ackrill and Kay (2011) illuminate a distinction between Kingdon's prediction of spillover across unconnected policy arenas (exogenous spillover) and spillover between institutionally connected areas (endogenous spillover). They argue that endogenous spillovers can hold windows of opportunity open, creating greater space for policy change.

2.2.5   Policy Entrepreneurs

Key to the coupling of the streams is the efforts of policy entrepreneurs. These uncommonly skilled individuals are adept at 'doing more with less' and packaging problems and solutions in a way such that they are able to respond appropriately when a policy window opens. While the European Commission is often identified as the pre-eminent EU policy entrepreneur, MSA assumes that the role may be played by any suitably equipped and skilled actor. Predictors of policy entrepreneurs' prospects for success include: access to key policymakers, the resources they have available to them and the strategies they use to couple the problem, policy and politics streams, as well as their political acumen. Entrepreneurs without formal institutional roles tend to rely on framing problems as a key strategy for influencing policy (Klüver et al. 2015). While entrepreneurship by civil society is necessary, but not sufficient on its own for such actors to wield influence over policy outcomes. In Kingdon's formulation, the specification of policy entrepreneurship offered by MSA is rather cursory and, since one objective of this study is to identify and describe policy entrepreneurs, Kingdon's specification appears inadequate to this task. To compensate, we add a supplementary specification of policy entrepreneurship, borrowing from Mintrom and Norman (2009) who set out four characteristics that can be used in order to identify policy entrepreneurs and entrepreneurship. These are social acuity, defining problems, building teams and leading by example (Mintrom and Norman 2009, p. 653).

2.3   Criticism and Application of MSA

Despite being apparently useful and highly cited, the MSA does have some critics. Kingdon's original work emerged from the 'garbage can' model organisational choice developed in the 1970s (Cohen et al. 1972), an approach which, while influential, was criticised on grounds of inconsistency and opacity (Bendor et al. 2001). The empirical testing and theoretical refinement of the original work, especially the improved specification of policy entrepreneurs, carried out by Kingdon and others, however, sets MSA apart from the original garbage can that inspired it.

Much criticism of MSA focuses on its underlying assumptions, most notably the assumption that the three streams (problem, policy and politics) are independent at all times other than during 'coupling' during a window of opportunity. Critics question this separation and argue that the streams are more interdependent than allowed for in MSA (Mucciaroni 1992; Bendor et al. 2001), a factor which puts MSA at odds with the ACF in which the three streams of activity are effectively fused (Sabatier and Jenkins-Smith 1993, 1999). In defence of the independence of the streams, Zahariadis (2014) argues that widespread empirical evidence for solutions being developed in the absence of problems strongly supports the conceptual device of stream independence. Indeed, the separation of the streams is likely to be reinforced in the context of this study by the fact that, almost by definition, interest group actors have a solution, in the form of a preferred policy, which they would like to couple with any appropriate problem.

The absence of clear and falsifiable hypotheses in the original formulation of MSA has been raised as a criticism, as has the somewhat vague nature of the language used to describe key concepts ('primaeval soup', for instance) (Herweg et al. 2018). While we contend that neither of these criticisms are especially pertinent to our study, it is important to observe that they are being effectively addressed by a recent resurgence of interest in this approach. It has been shown that falsifiable hypotheses can be built into MSA-led studies (e.g. Boscarino's (2009) hypotheses about problem surfing behaviour), although the absence of such general hypotheses from the approach does not hinder our use of it to explore a topic which essentially requires the telling of a story. It is also the case that improvements in the precision of the concepts are being made. For example, (Ackrill et al. 2013) add detail to the previously metaphorical description of 'spillover'. Increasing the precision of the concepts of MSA is also something that we return to in the conclusion.

2.3.1   *Application to the EU*

Although the origins of the multiple streams framework lie in the analysis of US Federal policymaking, the approach has been successfully applied to various facets of EU policymaking (Zahariadis 2008; Ackrill et al. 2013; Ackrill and Kay 2011). Moreover, the basic causal mechanisms of MSA, policy window opening and stream coupling, have been shown to be broadly applicable to EU energy policymaking (Herweg 2016). Additionally, MSA has been found to be especially useful in interpreting some of the EU's peculiarities, such as '*institutional fluidity, jurisdictional overlap, endemic political conflict, policy entrepreneurship and varying time cycles*' (Ackrill et al. 2013, p. 871; Herweg and Zahariadis 2018). As a result of this earlier work, there has been an upsurge in the number of studies using MSA to analyse EU policymaking, although many of these studies tend to make use of one of more conceptual tools from MSA rather than a full, systematic application of the framework (Herweg and Zahariadis 2018).

2.4   CONCLUSION

This chapter has introduced the MSA and made the case for its application to our chosen field of research. The chapter has also set out the principal assumptions that underpin the framework and the components that structure the primary empirical portion of this book. The core assumptions of MSA are:

(i) That policy is made in an environment of ambiguity and that preferences are problematic: Conditions that are prevalent in EU climate and energy policy fields;

(ii) That the time available to policymakers (and other actors) is scarce and that while the number of issues that can be attended to by an individual is small, as a community a much larger number can be considered; and

(iii) That the three streams (problem, policy and politics) flow independently through the policymaking process and that solutions (such as technology neutrality) may well pre-date the problems they are invoked to solve.

In conclusion, MSA is an adaptable framework that has been found not only suitable for studying the EU policy process, but in many ways is especially applicable to this particular policy setting, which has been described as an 'organised anarchy' (Herweg and Zahariadis 2018). The research reported in this volume seeks to exploit MSA to the greatest extent, in order to benefit from the insights the framework contains about nature of the policy process and to enable us to comment on the usefulness of the framework and future direction of MSA research. The following chapters now depict the three streams and their coupling in detail, drawing on contemporary documentary evidence such as consultations responses and policy proposals, as well as 32 élite interviews with Brussels policy experts.

REFERENCES

Ackrill, R., & Kay, A. (2011). Multiple Streams in EU Policy-Making: The Case of the 2005 Sugar Reform. *Journal of European Public Policy, 18*(1), 72–89.

Ackrill, R., Kay, A., & Zahariadis, N. (2013). Ambiguity, Multiple Streams, and EU Policy. *Journal of European Public Policy, 20*(6), 871–887.

Baumgartner, F. R., & Jones, B. (1991). Agenda Dynamics and Policy Subsystems. *The Journal of Politics, 53*(4), 1044–1074.

Bendor, J., Moe, T. M., & Shotts, K. W. (2001). Recycling the Garbage Can: An Assessment of the Research Program. *American Political Science Review, 95*(1), 169–190.

Börzel, T. A. (1998). Organizing Babylon: On the Different Conceptions of Policy Networks. *Public Administration, 76*(2), 253–273.

Boscarino, J. E. (2009). Surfing for Problems: Advocacy Group Strategy in U.S. Forestry Policy. *Policy Studies Journal, 37*(3), 415–434.

Cairney, P., & Jones, M. D. (2016). Kingdon's Multiple Streams Approach: What Is the Empirical Impact of this Universal Theory? *Policy Studies Journal, 44*(1), 37–58.

Cohen, M. D., March, J. G., & Olsen, J. P. (1972). A Garbage Can Model of Organizational Choice. *Administrative Science Quarterly, 17*(1), 1–25.

Haas, E. B. (1958). *The Uniting of Europe: Political, Social and Economic Forces, 1950–1957*. Stanford, CA: Stanford University Press (Reprint, Notre Dame, IN: University of Notre Dame Press, 2003).

Herweg, N. (2015). Against All Odds: The Liberalisation of the European Natural Gas Market—A Multiple Streams Perspective. In K. S. Jale Tosun & S. Schmitt (Eds.), *Energy Policy Making in the EU: Building the Agenda* (pp. 87–105). London: Springer.

Herweg, N. (2016). Explaining European Agenda-Setting Using the Multiple Streams Framework: The Case of European Natural Gas Regulation. *Policy Sciences, 49*(1), 13–33.

Herweg, N., Huß, C., & Zohlnhöfer, R. (2015). Straightening the Three Streams: Theorising Extensions of the Multiple Streams Framework. *European Journal of Political Research, 54*(3), 435–449.

Herweg, N., & Zahariadis, N. (2018). The Multiple Streams Approach. In N. Zahariadis & L. Buonanno (Eds.), *The Routledge Handbook of European Public Policy* (pp. 32–42). Oxford: Routledge.

Herweg, N., Zahariadis, N., & Zohlnhöfer, R. (2018). The Multiple Streams Framework: Foundations, Refinements and Empirical Applications. In C. M. Weible & P. A. Sabatier (Eds.), *Theories of the Policy Process*. Boulder, CO: Westview Press.

Howlett, M., Mcconnell, A., & Perl, A. (2016). Moving Policy Theory Forward: Connecting Multiple Stream and Advocacy Coalition Frameworks to Policy Cycle Models of Analysis. *Australian Journal of Public Administration, 76*(1), 1–15.

Jones, M. D., et al. (2016). A River Runs Through It: A Multiple Streams Meta-Review. *Policy Studies Journal, 44*(1), 13–36.

Kingdon, J. W., (2010). *Agendas, Alternatives, and Public Policies* (2nd ed.). Harlow: Pearson.

Klüver, H., Mahoney, C., & Opper, M. (2015). Framing in Context: How Interest Groups Employ Framing to Lobby the European Commission. *Journal of European Public Policy, 22*(4), 481–498.

Lorenzoni, I., & Benson, D. (2014). Radical Institutional Change in Environmental Governance: Explaining the Origins of the UK Climate Change Act 2008 Through Discursive and Streams Perspectives. *Global Environmental Change, 29,* 10–21.

McCool, D. (1998). The Subsystem Family of Concepts: A Critique and a Proposal. *Political Research Quarterly, 51*(2), 551–570.

Mintrom, M., & Norman, P. (2009). Policy Entrepreneurship and Policy Change. *Policy Studies Journal, 37*(4), 649–667.

Mucciaroni, G. (1992). The Garbage Can Model & the Study of Policy Making: A Critique. *Polity, 24*(3), 459–482.

Palmer, J. (2010). Stopping the Unstoppable? A Discursive-Institutionalist Analysis of Renewable Transport Fuel Policy. *Environment and Planning C: Government and Policy, 28*(6), 992–1010.

Princen, S. (2018). Agenda-Setting and Framing in Europe. *The Palgrave Handbook of Public Administration and Management in Europe* (pp. 535–551). London: Palgrave Macmillan.

Richardson, J. (2005). Policy-Making in the EU: Interests, Ideas and Garbage Cans of Primeval Soup. In J. Richardson (Ed.), *European Union: Power and Policy-Making* (pp. 3–30). New York: Routledge.

Rochefort, D. A., & Cobb, R. W. (1994). *The Politics of Problem Definition: Shaping the Policy Agenda*. Lawrence: University Press of Kansas.

Rozbicka, P., & Spohr, F. (2016). Interest Groups in Multiple Streams: Specifying Their Involvement in the Framework. *Policy Sciences, 49*(1), 55–69.

Sabatier, P. A. (1998, March). The Advocacy Coalition Framework: Revisions and Relevance for Europe. *Journal of European Public Policy, 5,* 98–130.

Sabatier, P. A. (2007). *Theories of the Policy Process* (2nd ed.). Boulder, CO: Westview Press.

Sabatier, P. A., & Jenkins-Smith, H. C. (1993). *Policy Change and Learning: An Advocacy Coalition Approach.* Boulder, CO: Westview Press.

Sabatier, P. A., & Jenkins-Smith, H. C. (1999). The Advocacy Coalition Framework. In P. A. Sabatier (Ed.), *Theories of the Policy Process* (pp. 117–166). Boulder, CO: Westview Press.

Tosun, J., Biesenbender, S., & Schulze, K. (2015). *Energy Policy Making in the EU: Building the Agenda.* London: Springer.

Weible, C. M., & Sabatier, P. A. (2017). *Theories of the Policy process* (4th ed.). London: Routledge.

Weible, C. M., & Schlager, E. (2016). The Multiple Streams Approach at the Theoretical and Empirical Crossroads: An Introduction to a Special Issue. *Policy Studies Journal, 44*(1), 5–12.

Winkel, G., & Leipold, S. (2016). Demolishing Dikes: Multiple Streams and Policy Discourse Analysis. *Policy Studies Journal, 44*(1), 108–129.

Zahariadis, N. (2007). Multiple Streams Framework: Structure, Prospects, Limitations. In P. A. Sabatier (Ed.), *Theories of the Policy Process* (pp. 65–92). Boulder, CO: Westview Press.

Zahariadis, N. (2008). Ambiguity and Choice in European Public Policy. *Journal of European Public Policy, 15*(4), 514–530.

Zahariadis, N. (2014). Ambiguity and Multiple Streams. In P. A. Sabatier & C. M. Weible (Eds.), *Theories of the Policy Process* (pp. 25–57). Boulder, CO: Westview Press.

CHAPTER 3

The Problem Stream

Abstract This chapter describes what Kingdon calls the 'problem stream'. The chapter sets out the debate surrounding the connected issues of 'energy' and 'climate' topics and outline the issues vying for European policymakers' attention in the year or so leading up to the European Commission's 2014 Communication on the Energy and Climate Framework for 2030. The conceivable list of potential problems relevant to the policy area may be extremely large but the list that actually receives attention is necessarily much shorter. The chapter focusses on problems of energy supply, environmental sustainability and the cost of energy.

Keywords Problem definition · Emissions trading · Energy security Renewable energy · Climate leadership · European Union

3.1 INTRODUCTION

This chapter identifies the EU climate and energy problems facing policymakers at the time the agenda was being set for the 2030 climate and energy framework. It provides an account of Kingdon's (2010) problem stream from which the problems that are deemed to require attention emerge. It is the first of three parallel chapters that describe each of the three streams of MSA.

© The Author(s) 2018   33
O. Fitch-Roy and J. Fairbrass, *Negotiating the EU's 2030 Climate and Energy Framework*, Progressive Energy Policy, https://doi.org/10.1007/978-3-319-90948-6_3

Conventionally, discussions about the problems that policymakers face with regard to energy are framed as a 'trilemma' in which three conflicting goals must be reconciled. These three elements typically are energy security, energy equity and environmental sustainability (World Energy Council 2015; Carbon Brief 2013; Falkner 2014). Similarly, the European Commission refers to three core objectives including security of supply and sustainability but 'equity' is replaced with 'competitiveness' (European Commission 2006). Competitiveness in this instance is a proxy for maintaining affordable energy supplies as well as the long-term Commission goal of completing the EU's internal energy market. The following subsections present the problems posed by each of the three objectives.

Section 3.2 of this chapter recaps and further develops the role and characteristics of the problem stream in MSA, Sect. 3.3 details the problems related to security of energy supply, Sect. 3.4 explores problems related to environmental sustainability and Sect. 3.5 looks at the problem of energy cost in Europe at a time of poor economic performance. Section 3.6 summarises and concludes the chapter.

3.2   MSA and the Problem Stream

According to Kingdon (2010), problems may become relevant for three main reasons. First, there may be a change in some more-or-less objective 'indicator'. Indicators, as seen by Kingdon, are not mere statements of the facts relevant to a problem. Often, different actors can interpret indicators in quite different ways as they seek to construct a problem to which they have an appropriate policy solution.

Second, problems may swing into view as a result of some 'focusing event' or crisis. A focusing event relevant to the energy and climate policy area, for example, may be a heightened focus on the cost of energy due to economic crisis unfolding during the negotiations. As with indicators, the means by which this mechanism brings problems to the attention of policymakers is not always straightforward. As well as making policy choices 'black and white' by making all other problems appear trivial, an event may be a 'symbol' for a subject that was in the minds of decision-makers already.

Finally, problems may become important due to perceptions of the performance of other policy decisions. In this instance, the policy for energy and climate for 2020 implemented in 2009 could provide such 'feedback' to actors in the policy subsystem about what works and what does not. Feedback may come from officially collected statistics[1] about the state of an earlier policy implementation, from complaints by the public-at-large or by a particular policymaker's personal experience of administering the policy.

As the following sections show, all of these mechanisms were active in bringing problems to the attention of policymakers during the negotiation of the 2030 framework.

3.3   Problems of Security of Energy Supply

Energy security is a notoriously slippery concept that evades neat definition (Chester 2010). Indeed, it has been argued that any analytical benefits gained from clear definition of the term do not necessarily translate into traction against real-world policy problems, at least partly because every actor in every system holds their own view about what 'energy security' may or may not mean (Mitchell and Watson 2013).

In the Brussels climate and energy policy community, the discussion of energy security problems tends to focus on one of two distinct concepts or meanings of the term, both related to the secure supply of energy. Most commonly referred to is the secure supply of primary energy products, especially when these are imported from non-EU countries and is often concerned with issues of geopolitics and reducing imports by securing greater domestic (EU) resource. The second definition emphasises a technical concern about the reliable operation of the electricity system and is especially concerned about questions of 'keeping the lights on' by ensuring that electricity supply is able to meet demand at all times.[2]

In this context, the former is captured by the term 'import dependence' and the latter by the term 'generation adequacy'. Each of these concepts is now discussed in turn. First, we examine Europe's chronic energy import dependence as well as the acute crisis that arose because of Ukraine and Russia's gas dispute. We then show how Europe's energy utilities, facing a threat to their business model, sought to frame

---

[1] From Eurostat, for example.

[2] Although generation adequacy is only one factor in ensuring reliable electricity supplies and others include frequency stability etc. (Cañizares et al. 2009).

renewable energy as a threat to electricity grid reliability and therefore energy security.

3.3.1   Import Dependence

The dependency of the European economy on imported energy sources is a perennial topic in EU energy policy discussions with interest 'spiking' from time to time. The supply of energy in the EU has undergone some significant shifts in recent decades. Production of formerly dominant fossil sources such as coal, oil and gas has waned while renewable energy sources' contribution to primary production has increased dramatically, especially in the last 10 years. However, the growth in renewables' contribution to total primary production has not been sufficient to offset the decline in fossil fuel production and total primary energy production in the EU fell between 1990 and 2014 (Eurostat 2016).

Consequently, to illustrate the point, it should be noted that more than half (53.2%) of the EU's gross inland consumption was imported from non-member countries in 2014 (Eurostat 2014). This proportion, known as the EU's 'energy dependence', varies very significantly by both energy source and member state although no EU member states are currently net exporters of energy.

Over the last decade, Russia has come to dominate the EU's supply of fossil fuel imports, accounting for approximately one-third of all imports of solid fuels, crude oil and natural gas. The exposure to a potential disruption to supplies from Russia to EU member states varies a great deal with the Baltic and some eastern European countries being especially vulnerable (Chyong and Tcherneva 2015).

By 2014, an anxiety that the EU's dependence on a single energy supplier would heighten the energy security risk had been part of the discourse about security of supply for many years (Dickel et al. 2014; Monaghan 2005; Helm 2005). This concern was underlined in 2006 and 2009 when gas supplies through Ukraine, an important route into the EU from Russia, accounting for about half of all Russian imports, were destabilised by a long-running contractual dispute between Russian and Ukrainian counter-parties (Łoskot-Strachota and Zachmann 2014; IEA 2016).

The dispute escalated once more in 2014 with the Russian annexation of Crimea on March 18, 2014. As part of the wider political crisis, gas supplies were again cut to Ukraine on 16 June 2014 leading to speculation about disrupted supplies to the EU the following winter, especially in the Baltic region and some Eastern European member states (Dickel et al. 2014). The result of the dispute in 2014 was to create a powerful focusing event that shaped policy actors' perceptions about the problems more or less at the same time as the 2030 energy and climate framework was being debated.

For example, the day after Ukraine's gas supply from Russia was interrupted, Ministers from seven member states[3] wrote to Commission President Barroso and Commissioners Oettinger and Hedegaard warning that:

> The current situation in the Ukraine emphasises the importance of reducing dependence on imported oil and natural gas. (Wathelet et al. 2014)

The issue of EU security of energy supply was framed in different ways by various actors and with varying degrees of success. In one example, at an informal meeting of EU energy ministers in Athens on 16th May 2014, the European Alliance of Companies for Energy Efficiency in Buildings (EuroACE) presented analysis aimed at drawing attention to the issue of European energy dependence (Greek Presidency of the Council of the European Union 2014; Joyce 2014). The presentation introduced the concept of 'energy dependence day': The theoretical point in the year after which Europe is '100% dependent' (EuroACE 2013). For example, the EU's energy dependence was 54%, thereby putting energy dependence day 46% of the way through the year or on the 18th June. This would be 38 days earlier than in 1995 when the EU's energy dependence was 43%,[4] demonstrating the increasing reliance on non-EU suppliers.

3.3.2   Electricity Generation Adequacy

The previous subsection looked at the problem of import dependence. This subsection focuses on related but separate issue of electricity generation adequacy.

---

[3] Belgium, Denmark, Germany, Greece, Ireland, Luxembourg, Portugal.

[4] Eurostat Code: tsdcc310.

In late 2007, the 20 largest European energy utility firms had a combined market capitalisation of an estimated €1 trillion. By the autumn of 2013, the companies' value had fallen by more than 50% (*The Economist* 2013). Germany's two largest utilities, EOn and RWE, lost more than three-quarters of their value during this period (Andresen 2014). A number of discrete events and ongoing trends were cited as being responsible for the fall in values these of firms. Among these were political factors such as the popular German decision in 2011, prompted by the nuclear disaster at Fukushima in Japan that year, to accelerate its withdrawal from nuclear energy in which several major utilities remained invested (Karnitschnig 2015). Other factors include the economic crisis that depressed economic activity and therefore electricity demand and a significant fall in the wholesale price of electricity, due in part to the displacement of conventional generation by zero-marginal-cost renewables (Andresen 2014).

As a result, in 2011/12 up to 20GW of gas-fired generation capacity was mothballed across Europe, much of it recent additions to the Utilities' fleets (Caldecott McDaniels and 2014). In 2013, the Magritte Group of Utility CEOs claimed that as much as 54GW of serviceable plant was unavailable to the Europe's electricity grids (Magritte Group 2013).

There was disagreement about both the causes and the implications of the overcapacity of gas-fired electricity plant. While some pointed to the utilities 'failure to adapt' to a changing world (Dallos 2014), the Magritte Group blamed it on new renewable capacity undermining existing generation and a carbon price too low to encourage coal-to-gas switching (meaning that the additions of zero-marginal-cost renewables pushed out gas, rather than coal). The group pointed out that, by 2013, significant conventional electricity generation capacity had been mothballed due to unprofitability. They put forward an argument which stated that the excessive subsidisation of new 'intermittent' renewable capacity was pushing out 'dispatchable' generation and thereby undermining Europe's ability to maintain the security of electricity supply. They argued that despite very high levels of generation capacity, variable renewables cannot be relied upon to provide energy when needed. Magritte Group founders, GDF Suez (now Engie) put it:

> Indeed, Europe has annual average power production overcapacity but lacks capacities to address consumption peaks. (GDF Suez 2013)

The idea that renewables undermine the security of electricity supply has a long heritage but was not universally accepted within the European Commission or the wider policy community. But, despite the availability of plenty of counter-arguments (For some examples, see Awerbuch 2006; Lund and Mathiesen 2009; Valentine 2011; Gardner et al. 2012; Fürstenwerth et al. 2015), the EU renewables sector did not engage directly with the specific problem of electricity network security. Instead, they tackled arguments about insecurity due to resource imports:

> ...better exploitation of the indigenous and unlimited renewable energy resources means a decrease in imports of ever more expensive fossil fuels. (Coalition of Progressive European Energy Companies quoted in Beckman 2013)

The result was that the Magritte Group utilities' argument that renewables growth resulting from the Renewable Energy Directive had become an electricity reliability problem went largely unchallenged.

3.4   PROBLEMS OF ENVIRONMENTAL SUSTAINABILITY

The previous section explored issues of security of energy supply from the perspectives of import dependence and electricity generation adequacy. This section explores the relevant problems of environmental sustainability facing the EU, which were:

- Europe's role as a global leader on climate change discussed in Sect. 3.4.1;
- the perceived risk that constraining emissions in the EU could encourage emissions elsewhere—so-called 'carbon leakage' discussed in Sect. 3.4.2; and
- the issue of low allowance prices in the EU-ETS covered in Sect. 3.4.3.

3.4.1   The Problem of EU Leadership on Climate Change

In October 2009, the Heads of State and Government of the EU set an objective for the EU[5] to reduce greenhouse gas emissions by 80–95% by

---

[5] And called on other developed countries to do the same.

2050—a political goal that has played a major role in defining the terms of climate and energy policymaking since (European Council 2009). Within the policy community, despite the misgivings of some industries, the decarbonisation objective is rarely explicitly questioned (with some notable exceptions, e.g. Euracoal 2014) and provides the overarching frame in which discussion is conducted.

The EU has played a leading role in the international climate negotiation sphere since the 1990s (Oberthür and Roche Kelly 2008; Oberthür 2011). However, following a poor leadership performance at the COP 15 conference in Copenhagen, the ability of the EU to claim leadership status was called into question (Groen et al. 2012; Oberthür 2011). It enabled actors interested in a less ambitious approach, including BUSINESSEUROPE and some central and eastern European member states, to make the case that, as one business lobbyist put it in interview:

> "...no-one is doing anything similar across the rest of the world"..."[the EU] is a leader with no followers". (Interview 29, 2016)

3.4.2   The Problems of 'Carbon Leakage' and Deindustrialisation

Having won a concession, in the form of a 'carbon leakage list' of sectors which are deemed to be at particular risk of being forced to relocation outside of the EU by climate policy costs and therefore eligible for free emissions allowances (Juergens et al. 2013), the energy-intensive sectors continued throughout 2013–2014 to noisily draw attention to what is known as the 'carbon leakage' issue—alongside the threat of deindustrialisation—as the major problem faced by EU climate and energy policy (IFIEC 2014).

The perceived potential for the unintended increase of (GHG) emissions outside the EU due to constrained emissions inside the EU has been part of the EU climate and energy conversation since at least the launch of the ETS in 2005 (Babiker 2005).

Carbon leakage is generally taken to refer to either the relocation of industrial plants to, or an increase in imports from, countries with less stringent emissions constraints. One of the earliest uses of the term was by the alliance of energy-intensive industries in a contribution to the Commission's stakeholder engagement process leading up to a revision of the ETS Directive in 2008 (The Key Stakeholders Alliance for ETS

Review 2007; Koch and Mama 2016). The cost impact and the ability to pass those costs along the supply chain (typically to consumers) are the two major risk factors determining the risk of carbon leakage from a particular industry, clearly distinguishing between the electricity generation sector which is largely able to pass costs onto consumers from the manufacturing sectors which face competition from non-EU suppliers (Marcu et al. 2013). Concerns about the impact of climate policy on European industry's global competitiveness are compounded by the fact that industry's share of European Gross Domestic Product (GDP) has been continually declining for 25 years (European Commission 2013b).

3.4.3   The Problems with EU-ETS

During 2012 and 2013, a very large oversupply of allowances developed and the Emission Trading System (EU-ETS) price fell to well below €10 per tonne, a price that most agreed was too low to have a meaningful impact on investment decisions in the energy sector (e.g. EWEA 2014; European Commission 2014). As in an indicator, the price was taken by many to be a clear demonstration that the EU-ETS was unable to perform as an effective instrument.

To give an impression of how widespread the sense that the EU-ETS was failing had become, one Commission official in DG Energy stated in an interview that:

> [The] ETS price has never driven any investment, not even the management of existing assets from a company, the CO2 price has never driven the running or non-running of a renewables versus coal power plant of a given portfolio of a company. (Interview 7, 2015)

And the chief climate change advisor for Shell, a firm supporter of a 'strong, reformed ETS' conceded that:

> …many now perceive that the EU-ETS has become more of a compliance formality than an investment driver. (Hone 2015)

The fact that the use of coal, the most emissions-intensive electricity generation fuel, actually increased in some member states in 2013 (especially Germany) served to underline the failure of the trading system to influence decision-making (Platts 2013).

ETS Reform Overlapping with 2030

For the reasons set out above, during the period in which the 2030 framework was under discussion, reform of the ETS was also being negotiated within the Brussels climate and energy policy community. The reform options pursued by the Commission included 'backloading' or temporarily withholding allowances scheduled to be auctioned to the market, a proposal that was adopted by the Parliament in July 2013 and the Council in December the same year.[6] Second, the Commission proposed structural ETS reform in the form of a 'Market Stability Reserve' (MSR), a mechanism that allows the number of permits issued to the market to vary according to the number of allowances in circulation. The MSR proposal put forward by the Commission was adopted by the Parliament in July 2015 and the Council in October of the same year (European Commission 2014c; European Parliament 2014a). The MSR is expected to enter into operation in 2019.

The overlap between the debate about ETS reform and the 2030 targets had a significant impact on actors' strategies as well as the way in which they framed the problem to be targeted by the 2030 framework, which will be discussed in more detail in Chapter 6.

Framing and Feedback

The way in which the various actors prioritise or omit the different explanations in their framing of the problem represented by the collapse in the carbon price varied markedly. While very nearly all actors acknowledged that the economic crisis had a significant role in suppressing the ETS price and therefore the policy's ability to drive low-carbon investment, there was widespread disagreement, first about which other factors were important and, second, about the relative significance of these other factors.

From industry, the large energy utilities, oil companies, energy-intensive industries and the carbon trading industry tended to put forward an argument contending that, second to the economic crisis, the major factor explaining the low price of carbon in the EU was the impact of additional carbon savings from alternative climate policies such as the 2009 renewable energy and 2012 energy efficiency directives. Some, such as a coalition including Shell and Fortum and Areva argued that, in

---

[6] Decision No 1359/2013/EU.

the early trading periods it had been international credits adding to supply but that in the most recent phase, the most important single factor in explaining the oversupply of EUAs was the 'overlap' between the ETS and the 2020 targets (Interview 15, 2015; Interview 25, 2016). Some utility actors made a case that the low ETS price was not a problem in and of itself and it was the independent renewables target continuing to drive deployment at a time of low price that was the problem. The energy efficiency community went to great efforts to show that energy efficiency does not necessarily undermine carbon emissions reduction, as stated by some producer industries (The Coalition for Energy Savings 2013; Eichhammer 2013).

While some environmental NGOs acknowledged that overlapping policies may have a role to play, some described it as '*the smallest problem facing the EU's cap-and-trade system*' (Sandbag 2013). Many of the Brussels-based NGOs acknowledged that a very low ETS price may be problematic for policymakers but did not actively diagnose its cause. One NGO, however, described blaming energy efficiency measures for undermining the carbon price as 'twisted logic' making the case that '*the point is that emissions are cut, not how they are cut*' (Friends of the Earth 2013). Many NGOs are explicit in their assertion that the free allocation of allowances in the first two phases of the ETS led to over-allocation to industry and the ability to 'grandfather' or carry-over allowances from one trading period to the next was to blame for the low prices. Interviewees identify the cause as both a basic design flaw of the ETS and the result of successful lobbying by industry. There was also some concern that the free allowance of allocations had resulted in significant windfalls for some industries (Laing et al. 2013).

Among member states, the UK pushed its position in support of the principle of 'technology neutrality' in EU energy policy by citing a range of causes including policy interaction. It also stressed the lack of ambition in the 20% target being an important factor (DECC 2013). Conversely, Poland stated that the low prices in the ETS were not necessarily a problem, rather a natural outcome of the economic downturn (Government of Poland 2013).

In the documents supporting the Commission's proposals for 2030 energy and climate policy, there is very little diagnosis of the low prices (European Commission 2013a, 2014b). Some utilities point out that the ETS did exactly what it ought to do in a recession: That is to reduce the price of allowances (Vattenfall 2013).

There is some tension within the European Commission about the causes of low ETS prices. For example, DG Clima, the Directorate-general responsible for the design and reform of the ETS raised concerns that renewable energy or energy efficiency policy may act to drive down prices in 2011 (van Renssen 2014b), a position generally at odds with that taken by people working in DG Energy (Interview 17, 2015; Interview 28, 2016).

Some actors in Brussels sought to frame nearly all climate and energy challenges facing the EU as an ETS problem. One campaign, known as the 'Friends of ETS' (FoETS), expended a great deal of effort emphasising the 'crisis' in the ETS. The primary goal of the campaign was to ensure the passage of two ETS reform packages, backloading and the market stability reserve through Parliament. One strategy used in this campaign was to attach questions of ETS reform to all relevant policy debates in Brussels by emphasising crisis in the system. In the words of one participant in the campaign:

> "So we started to have huge surpluses, the price is falling again; we have crisis, crisis, crisis"… "… you've always got to have a permanent state of crisis."…"…So every single bit of coverage leading up to any communication from the Commission [*on energy and climate*] we said, "We have a crisis in the ETS"". (Interview 30, 2016)

In the run-up to the October 2012 publication of the Energy Efficiency Directive implementing the 2020 target, 'policy overlap' was invoked by the same coalition as a reason for the Commission to focus attention on the ETS:

> [*We knew*] the energy efficiency debate meant that [energy efficiency] was getting all the coverage. [*So we made sure that*] every time somebody spoke about energy efficiency, the whole discussion was about ETS. (Interview 30, 2016)

To underline the crisis in the ETS, the group also pointed to existing and proposed policies:

> So we started [*saying*], 'The Energy Efficiency Directive will destroy the ETS, because it will create additional surplus.' And maybe it will do, later on, after… Assuming all the energy efficiency does happen, but there's a time lag before that happens. But we said up front 'it's about 400 million

tonnes worth the CO2's going to be taken out of the market, that's going to kill the ETS, it's a disaster, we have to do something!'. (Interview 30, 2016)

And, indeed, an acknowledgement that the two policies were inter-linked did make its way into the final text of the Energy 2012 Efficiency Directive.

3.5   THE PROBLEM OF THE COST OF ENERGY IN RECESSION EUROPE

This section considers the implications of rising energy prices in the context of weak economic performance. First, the nature and scale of the focussing event of the economic crisis is set out in Sect. 3.5.1. Second, we explore two associated problems. In the 2030 discussion, concerns about the absolute and relative cost of energy, sometimes described as its 'affordability' can be divided into two main categories: The impact on homes and businesses of changes to energy costs, discussed in Sect. 3.5.2 and the widespread perception that supporting renewable energy technologies was a contributing factor to rising costs discussed in Sect. 3.5.3.

3.5.1   Economic Crisis

The single most important event that occurred between the negotiation of the 20/20/20 package in the 2007 and the discussion about a post-2020 package is the ongoing financial crises and European economic recession. Although Europe was officially out of recession by 2013, issues of economic growth and employment dominated all areas of European political life. In some ways, there may have been some benefits for ambitious climate policy:

> So what we were saying is kind of we know we need to get to [80% decar-bonisation by] 2050, we're in the midst of a recession, in some ways a recession is helping us, because obviously we're consuming less, we're decoupling from greenhouse gas emissions from growth. (Interview 29, 2016)

But it was not, of course, all good news. Investment and employment were both weak and European industries were taking an unexpectedly long time to recover. Such recovery that there was seemed weaker than many of the EU's global trading partners, worrying policymakers who blamed the situation, in part, on high energy prices (European Commission 2013b; European Commission 2014).

The difficult economic situation had a significant impact on the framing of the problems facing climate and energy policy with industry seeking to emphasise the state of crisis:

> ...but then on the other hand it's hitting our industries...our core industries very hard. And their intense push-back is really starting to be felt across the Commission. (Interview 29, 2016)

> Don't underestimate [*that*] in an economic crisis, policymakers are destabilised and it's the best moment ever for business people. (Interview 4, 2015)

Feedback from the previously implemented 2020 targets also became a key driver of the direction of the debate about the post-2020 policy:

> In terms of why there has been a change in direction [*since 2007*] - it's the economy, stupid. The cost effectiveness of any policy has become more important and as Europe has gone through this economic crisis and slump. I think there is no question that [the case can be made that] the 2020 targets cost a lot of money. (Interview 25, 2015)

3.5.2   Focus on the Cost of Energy

In light of the ongoing financial crisis, the cost of energy to homes and businesses in the EU in 2013 and 2014 was a major challenge. According to a report requested by the European Council in May 2013 and published by the Commission in January 2014, the price of natural gas and electricity showed that average EU household electricity prices had risen 4% per year between 2008 and 2012 and gas prices 3% per year. Both were faster than inflation. Industrial retail energy prices had risen less sharply with electricity increasing by 3.5% per year and gas by less than 1% per year (European Commission 2014a).

The perception that overall energy costs in the EU appeared stubbornly high contrasted a long-term decline in gas prices in the USA, a major trading partner for the EU. This change in the US gas market was largely the result of large-scale exploitation of domestic shale gas deposits.

Between 2000 and 2010, shale gas expanded its share of US domestic gas production from less than 1% to more than a 20%, leading to a very substantial oversupply and consequent price falls (Stevens 2012).

As well as the implications for global competitiveness, the so-called 'shale gas revolution' in the USA led to an increase in US coal exports, contributing to a reduction in global coal prices during the period in which the 2030 targets were being discussed (European Commission 2014a; Cornot-Gandolphe 2015). Declining coal prices, increased concerns that coal—a very emissions-intensive source of energy—would be harder than hoped to eradicate from the European fuel mix (Zachmann 2015; Chazan and Wiesmann 2013).

The potential impact on the global competitiveness of the struggling EU economy and the changes to coal and gas prices were fundamental to the terms of the debate about the 2030 targets. In the words of one participant:

> I would say that the economy and the shale gas revolution in the US fundamentally changed the dynamics. It's the reason why we have cheap coal. It's the reason EU competitiveness is suffering. These macro pictures had a big impact [*on the debate*]. (Interview 25, 2015)

The increase in energy prices, especially for households, was largely due to taxes, levies for specific policies and network costs. Some of these costs are associated with climate and energy policies such as supporting renewable energy investment.[7] Supporting renewable energy accounted for 6% of the average EU household electricity price and 8% of the industrial electricity price in 2013. However, there is significant variation among member states in the proportion of electricity prices made up by renewable energy support costs. For example, in Spain and Germany the share was more than 15% while in Ireland, Poland and Sweden it was less than 1%.

3.5.3   The Cost of Supporting Renewable Energy

In 2013 and 2014, the heightened concern about the cost of energy led to a focus on the cost of supporting renewable energy in the EU. For at least a decade, the most commonly used policy instrument for supporting renewable electricity has been the feed-in-tariff (FIT)

---

[7] EU-ETS costs are passed into wholesale energy costs.

(Kitzing et al. 2012). The feed-in instrument, which offers revenue certainty for projects, has been shown to be a very effective means of promoting renewable electricity expansion compared to the most commonly used alternative, the tradable quota system (Mitchell et al. 2006; Ringel 2006). But, this success, especially as the cost of renewables such as solar photovoltaics (PV) fell rapidly, had come at a cost.

In some EU countries, the rate of growth of some renewable energy sources, and the associated cost of remuneration, became politically problematic. In Germany, for example, in pursuit of the country's 2020 target, seven gigawatts of solar capacity were added to the FIT per year between 2010 and 2013, amounting to an addition of more than €10bn to 2014 energy bills through a surcharge (Weiss 2014). While Germany acted in 2014 to limit the cost of future implementation (BMWi 2014), very little could be done to reduce the impact of the costs of electricity bills save reducing some of the pre-existing exemptions for industry to reduce the burden on households. Spain is another member state that used FITs very successfully. As in Germany, the cost of supporting renewables in Spain was seen as politically challenging, especially following the financial crisis, leading to, from 2010, a series of changes to remuneration levels for new and existing projects and eventually a moratorium on new renewable plants (del Río and Mir-Artigues 2012).

The concern over 'run-away' renewables costs and its impact on business, industrial and domestic energy consumers was taken up at a European level by the 'Magritte Group' of utility CEOs in 2013 which referred to events in Germany and Spain to call attention to the impact on electricity prices and markets[8] of large volumes of zero-marginal-cost generation (Magritte Group 2013).

In June 2014, the European Commission published new guidelines on conditions under which government support for renewables would be exempted from some elements of European competition law. The new 'state-aid' guidelines required that all member states refrain from the use of instruments such as feed-in tariffs and instead use 'competitive bidding processes' such as auctions to allocate support to renewables, an approach that enables governments to control the volume of deployment of renewables much more tightly while at the same time promoting

---

[8] Through the so-called 'merit order effect' (see Sensfuß et al. 2008 for more detail).

competition in the renewables sector in the hope of minimising electricity costs (Fitch-Roy 2016; Winkler et al. 2018).

3.6   Conclusion

MSA suggests that agenda-setting takes place with respect to specific policy problems and that the problems being discussed and their relative priority depends on indicators, focussing events and feedback about problems by political actors. This chapter has described what Kingdon (2010) calls the problem stream in which problems are identified.

In spite of the fact that the EU's global leadership role had been seen to slip following COP 15 in Copenhagen, the primary problem that set the terms of discussion in the climate and energy policy community was that of decarbonising the European economy by 2050, a timetable set by the European Council in 2009.

However, the flagship climate policy, the EU-ETS was widely accepted to be failing in its task of creating an incentive to reduce emissions, as indicated by very low emissions prices. The 'crisis in the ETS' was heavily emphasised by some environmental and business interests in order to secure ETS reform, a process which spilt over into questions about the interaction with other policies. In particular, the role that renewable energy and energy efficiency policy could play in undermining the carbon price. This framing of the ETS problem as a renewables and energy efficiency problem was emphasised by actors such as the Magritte Group and the 'single-target coalition'. Even some environmental actors, attempting to ensure ETS reform, actively promoted an argument that pitted renewables and energy efficiency against the ETS.

The energy-intensive industries and BUSINESSEUROPE continued to make the case that the cost of climate policy on industry posed a major problem to European competitiveness, industrial base and employment, a position they have held since the inception of climate policy and carbon trading.

The pall of Europe's poor economic and industrial performance hung over European political life throughout the agenda-setting phase of the post-2020 energy and climate policy. The economic crises, indicated by economic statistics, focussed attention on industrial competitiveness and the cost of policy with experience of renewables support cost from member states such as Germany and Spain often pointed out as salutary reminders of what can happen when policymakers 'lose control' of policy costs.

The political crisis in Ukraine and the threat of disrupted gas supplies powerfully focussed attention on Europe's energy dependence and wider issues of security of energy supply. The energy efficiency lobby led by EuroACE were able to associate themselves with the issue through the creation of a useful heuristic tool, 'energy dependence day'. Meanwhile, the Magritte group and others were somewhat successful in making a case that the rapid rise of renewables such as wind and solar due to the 2009 Renewable Energy Directive was pushing more reliable gas plant out of the market and therefore putting the reliability of some European electricity systems in jeopardy. The renewable energy industry did not forcefully oppose this feedback.

This chapter has established which problems were pertinent to the debate about the 2030 targets and shows that interest groups made use of indicators such as the prices of emissions allowances and economic performance, focussing events such as the Ukraine politics crisis, and feedback about the 20/20/20 package of policies to frame particular problems as worthy of attention.

MSA proposes that, in order to influence policy outcomes, actors must match problems with viable solutions to those problems. The next chapter describes the range of solutions that emerge from what Kingdon (2010) calls the policy stream.

REFERENCES

Andresen, T. (2014). *German Utilities Hammered in Market Favoring Renewables—Bloomberg Business*. Bloomberg.com. Available at: http://www.bloomberg.com/news/articles/2013-08-11/german-utilities-hammered-in-market-favoring-renewables. Accessed 1 Dec 2015.

Awerbuch, S. (2006). Portfolio-Based Electricity Generation Planning: Policy Implications for Renewables and Energy Security. *Mitigation and Adaptation Strategies for Global Change, 11*(3), 693–710.

Babiker, M. H. (2005). Climate Change Policy, Market Structure, and Carbon Leakage. *Journal of International Economics, 65*(2), 421–445.

Beckman, K. (2013). *'Progressive Energy Companies' Versus Magritte Group*. Available at: http://www.energypost.eu/progressive-energy-companies-versus-margritte-group/. Accessed 1 Feb 2014.

BMWi. (2014). *EEG 2014*. Available at: http://www.bmwi.de/DE/Themen/Energie/Erneuerbare-Energien/eeg-2014.html. Accessed 31 July 2016.

Caldecott, B., & McDaniels, J. (2014). *Stranded Generation Assets: Implications for European Capacity Mechanisms, Energy Markets and Climate Policy Working Paper.* Available at: http://www.smithschool.ox.ac.uk/research/sustainable-finance/publications/Stranded-Generation-Assets.pdf. Accessed 10 Mar 2014.

Cañizares, C., Rouco, L., & Andersson, G. (2009). Angle, Voltage and Frequency Stability. In A. Gomez-Exposito, A. J. Conejo, & C. Canizares (Eds.), *Electric Energy Systems: Analysis and Operation.* Boca Raton: CRC Press.

Carbon Brief. (2013). *Climate Rhetoric: What's an Energy Trilemma?* Available at: https://www.carbonbrief.org/climate-rhetoric-whats-an-energy-trilemma. Accessed 10 Oct 2016.

Chazan, G., & Wiesmann, G. (2013). *Shale Gas Boom Sparks EU Coal Revival.* FT.com. Available at: https://www.ft.com/content/d41c2e8a-6c8d-11e2-953f-00144feab49a. Accessed 10 Oct 2016.

Chester, L. (2010). Conceptualising Energy Security and Making Explicit Its Polysemic Nature. *Energy Policy, 38*(2), 887–895.

Chyong, C.-K., & Tcherneva, V. (2015). *Europe's Vulnerability on Russian Gas.* Available at: http://www.ecfr.eu/article/commentary_europes_vulnerability_on_russian_gas. Accessed 1 Nov 2016.

Cornot-Gandolphe, S. (2015). *US Coal Exports: The Long Road to Asian Markets.* Available at: https://www.oxfordenergy.org/wpcms/wp-content/uploads/2015/03/CL-21.pdf. Accessed 28 Nov 2016.

Dallos, G. (2014). *Locked in the Past: Why Europe's Big Energy Companies Fear Change.* Hamburg: Greenpeace.

DECC. (2013). *2030 Green Paper Response.* Available at: https://www.gov.uk/government/uploads/system/uploads/attachment_data/file/210659/130703_response_for_publication.pdf. Accessed 10 Feb 2014.

Dickel, R., et al. (2014). *Reducing European Dependence on Russian Gas: Distinguishing Natural Gas Security from Geopolitics.* Oxford: Oxford Institute of Energy Studies.

Eichhammer, W. (2013). *Analysis of a European Reference Target System for 2030. Energy Savings 2030: On the 2050 Pathway.* Available at: http://www.isi.fraunhofer.de/isi-wAssets/docs/x/de/publikationen/Fraunhofer-ISI_ReferenceTargetSystemReport.pdf. Accessed 11 Feb 2016.

Euracoal. (2014). *Why Less Climate Ambition Would Deliver More for the EU.* Brussels.

EuroACE. (2013). *EuroACE Position Paper on EU Post 2020 Climate and Energy Policy: EuroACE Supports a Binding 2030 Energy Efficiency Target.* Brussels.

European Commission. (2006). *A European Strategy for Sustainable, Competitive and Secure Energy.* Available at: http://europa.eu/documents/comm/green_papers/pdf/com2006_105_en.pdf. Accessed 14 Apr 2014.

European Commission. (2013a). Draft Commission Staff Working Document Impact Assessment for a 2030 Climate and Energy Policy Framework.

European Commission. (2013b). *Member States' Competitiveness Performance and Implementation of EU Industrial Policy.* Available at: http://ec.europa. eu/DocsRoom/documents/110/attachments/1/translations/en/renditions/native. Accessed 17 Feb 2016.

European Commission. (2014). *Energy prices and costs in Europe.* Available at: https://ec.europa.eu/energy/sites/ener/files/documents/20140122_communication_energy_prices.pdf. Accessed 7 Feb 2014.

European Commission. (2014a). *Energy Prices and Costs in Europe.* Accessed 7 Feb 2014.

European Commission. (2014b). *Impact Assessment—A Policy Framework for Climate and Energy in the Period from 2020 up to 2030.* Available at: http://ec.europa.eu/smart-regulation/impact/ia_carried_out/docs/ia_2014/swd_2014_0015_en.pdf. Accessed 13 July 2015.

European Commission. (2014c). *Proposal for the Establishment and Operation of a Market Stability Reserve for the Union Greenhouse Gas Emission Trading Scheme and Amending Directive 2003/87/EC.* Available at: http://eur-lex.europa.eu/legal-content/EN/TXT/PDF/?uri=CELEX:52014PC0020&from=en. Accessed 24 Nov 2015.

European Council. (2009). *29/30 October 2009: Conclusions.* Available at: http://www.consilium.europa.eu/uedocs/cms_data/docs/pressdata/en/ec/110889.pdf. Accessed 31 Mar 2016.

European Parliament. (2014a). *Decision of the European Parliament and of the Council Concerning the Establishment and Operation of a Market Stability Reserve for the Union Greenhouse Gas Emission Trading Scheme and Amending Directive 2003/87/EC.* Brussels.

Eurostat. (2014). *Energy Balance Sheets 2011–2012.* Available at: http://ec.europa.eu/eurostat/documents/3217494/5785109/KS-EN-14-001-EN.PDF/16c0ac97-7dd6-4694-b22d-e77a36cb4e86. Accessed 10 Feb 2016.

Eurostat. (2016). *Primary Production of Energy by Resource.* Available at: http://ec.europa.eu/eurostat/tgm/table.do?tab=table&init=1&language=en&pcode=ten00076&plugin=1. Accessed 22 Feb 2018.

EWEA. (2014). *5 Priorities for a European Energy Union.* Available at: http://www.ewea.org/fileadmin/files/library/publications/position-papers/EWEA-EU-Energy-Union-5-Priorities.pdf. Accessed 27 Apr 2018.

Falkner, R. (2014). Global Environmental Politics and Energy: Mapping the Research Agenda. *Energy Research & Social Science, 1,* 188–197.

Fitch-Roy, O. W. (2016). An Offshore Wind Union? Diversity and Convergence in European Offshore Wind Governance. *Climate Policy, 16*(5), 586–605.

Friends of the Earth. (2013). *Comments on the EC Green Paper 'A 2030 Framework for Climate and Energy Policies'.* Brussels.

Fürstenwerth, D., Pescia, D., & Litz, P. (2015). *The Integration Costs of Wind and Solar Power.* Agora Energiewende. Available at: https://www.agora-ener-giewende.de/fileadmin/Projekte/2014/integrationskosten-wind-pv/Agora_Integration_Cost_Wind_PV_web.pdf. Accessed 18 Feb 2016.

Gardner, P., Fitch-Roy, O. W., & Platt, R. (2012). *Beyond the Bluster: Why Wind Power Is an Effective Technology.* London: IPPR.

GDF Suez. (2013). *Consultation on a 2030 Framework for Climate and Energy Policies GDF SUEZ Answer.* Available at: http://ec.europa.eu/energy/con-sultations/doc/com_2013_0169_green_paper_2030_en.pdf. Accessed 19 Feb 2016.

Government of Poland. (2013). *2030 Green Paper Response.* Brussels.

Greek Presidency of the Council of the European Union. (2014). *Informal Meeting of Energy Ministers Athens, 15–16 May 2014 'Energy Security' Discussion Paper.* Available at: http://gr2014.eu/sites/default/files/DiscussionPaperonEnergySecurity.pdf. Accessed 27 Apr 2016.

Groen, L., Niemann, A., & Oberthür, S. (2012). The EU as a Global Leader? The Copenhagen and Cancun UN Climate Change Negotiations, *8*(2), 173–191.

Helm, D. (2005). *European Energy Policy: Securing Supplies and Meeting the Challenge of Climate Change.* Oxford.

Hone, D. (2015). *Putting the Genie Back: Why Carbon Pricing Matters.* Whitefox.

IEA. (2016). *Gas Trade Flow in Europe.* Available at: https://www.iea.org/gtf/. Accessed 11 Apr 2016.

IFIEC. (2014). *Manifesto: Europe's Manufacturing Industry CEOs Call Upon Heads of State to Streamline 2030 Strategy Towards Growth and Jobs.* Brussels.

Joyce, A. (2014). *Speech to the Informal Energy Ministers Meetings—Athens—16th May 2014: Financing of Energy Efficiency Measures.* Athens.

Juergens, I., Barreiro-Hurlé, J., & Vasa, A. (2013). Identifying Carbon Leakage Sectors in the EU ETS and Implications of Results. *Climate Policy, 13*(1), 89–109.

Karnitschnig, M. (2015). *Germany's Green Power Meltdown.* politico.eu. Available at: https://www.politico.eu/article/germanys-green-power-melt-down/. Accessed 22 Feb 2018.

Kingdon, J. W. (2010). *Agendas, Alternatives, and Public Policies* (2nd ed.). Harlow: Pearson.

Kitzing, L., Mitchell, C., & Morthorst, P. E. (2012). Renewable Energy Policies in Europe: Converging or Diverging? *Energy Policy, 51,* 192–201.

Koch, N., & Mama, H. B. (2016). *European Climate Policy and Industrial Relocation: Evidence from German Multinational Firms.* Available at: https://papers.ssrn.com/sol3/Delivery.cfm/SSRN_ID2868283_code1302307.pdf?abstractid=2868283&mirid=1. Accessed 5 Dec 2016.

Laing, T. et al. (2013). *Assessing the Effectiveness of the EU Emissions Trading System* (CCCEP Working Paper, 126).

Łoskot-Strachota, A., & Zachmann, G. (2014). *Rebalancing the EU-Russia-Ukraine Gas relationship.* Available at: https://www.econstor.eu/dspace/bitstream/10419/106321/1/812740270.pdf. Accessed 11 Feb 2016.

Lund, H., & Mathiesen, B. V. (2009). Energy System Analysis of 100% Renewable Energy Systems—The Case of Denmark in Years 2030 and 2050. *Energy, 34*(5), 524–531.

Magritte Group. (2013). *Press Release: Heads of 12 Leading European Energy Companies Propose Concrete Measures to Rebuild Europe's Energy Policy.* Available at: https://www.engie.com/wp-content/uploads/2013/11/12CEO_VA_v4.pdf. Accessed 19 Feb 2014.

Marcu, A. et al. (2013). *Carbon Leakage: An Overview.* Available at: https://www.ceps.eu/system/files/SpecialReportNo79CarbonLeakage_0.pdf. Accessed 11 Feb 2016.

Mitchell, C., Bauknecht, D., & Connor, P. M. (2006). Effectiveness Through Risk Reduction: A Comparison of the Renewable Obligation in England and Wales and the Feed-in System in Germany. *Energy Policy, 34*(3), 297–305.

Mitchell, C., & Watson, J. (2013). Introduction: Conceptualising Energy Security. In C. Mitchell, J. Watson, & J. Britton (Eds.), *New Challenges in Energy Security: The UK in a Multipolar World* (pp. 1–21). Basingstoke: Palgrave Macmillan.

Monaghan, A. (2005). *Russian Oil and EU Energy Security.* Conflict. Available at: https://www.files.ethz.ch/isn/96125/05_Nov.pdf. Accessed 19 Feb 2016.

Oberthür, S. (2011). The European Union's Performance in the International Climate Change Regime. *Journal of European Integration, 33,* 667–682.

Oberthür, S., & Roche Kelly, C. (2008). EU Leadership in International Climate Policy: Achievements and Challenges. *The International Spectator, 43*(3), 35–50.

Platts. (2013). *German Coal-Fired Power Rises Above 50% in First-Half 2013 Generation Mix—Electric Power | Platts News Article & Story.* Available at: http://www.platts.com/latest-news/electric-power/london/german-coal-fired-power-rises-above-50-in-first-26089429. Accessed 18 Feb 2016.

van Renssen, S. (2014b). *Split Emerges in the Commission Over Energy-Efficiency Measures.* Politico. Available at: http://www.politico.eu/article/split-emerges-in-the-commission-over-energy-efficiency-measures/. Accessed 8 May 2014.

Ringel, M. (2006). Fostering the Use of Renewable Energies in the European Union: The Race Between Feed-in Tariffs and Green Certificates. *Renewable Energy, 31*(1), 1–17.

del Río, P., & Mir-Artigues, P. (2012). Support for Solar PV Deployment in Spain: Some Policy Lessons. *Renewable and Sustainable Energy Reviews, 16*(8), 5557–5566.

Sandbag. (2013). *Consultation Response—2030 Energy and Climate Framework Green Paper.* Available at: https://crowdsourcing.simpolproject.eu/static/staticdata/gpc/consultations/sandbag.pdf. Accessed 30 Nov 2015.

Sensfuß, F., Ragwitz, M., & Genoese, M. (2008). The Merit-Order Effect: A Detailed Analysis of the Price Effect of Renewable Electricity Generation on Spot Market Prices in Germany. *Energy Policy, 36*(8), 3076–3084.

Stevens, P. (2012). *The 'Shale Gas Revolution': Developments and Changes.* London: Chatham House.

The Coalition for Energy Savings. (2013). *A Binding Energy Savings Target for 2030: The Cornerstone for Mutually Supporting Climate and Energy Policies.* Available at: https://www.eurima.org/uploads/ModuleXtender/Publications/105/20131011_Coalition_position_on_2030.pdf. Accessed 7 Apr 2016.

The Economist. (2013). European Utilities: How to Lose Half a Trillion Euros. *The Economist.* Available at: http://www.economist.com/news/briefing/21587782-europes-electricity-providers-face-existential-threat-how-lose-half-trillion-euros. Accessed 16 Apr 2014.

The Key Stakeholders Alliance for ETS Review. (2007). *Lowering Production Is No Benefit for the Environment, Says European Industry.* Brussels.

Valentine, S. V. (2011). Emerging Symbiosis: Renewable Energy and Energy Security. *Renewable and Sustainable Energy Reviews, 15*(9), 4572–4578.

Vattenfall. (2013). *2030 Green Paper Response.* Brussels.

Wathelet, M. et al. (2014, June 17). Letter Calling for a Proposal on a Binding Energy Efficiency Target for 2030.

Weiss, J. (2014). *Solar Energy Support in Germany: A Closer Look.* Washington, DC: Brattle.

Winkler, J., Magosch, M., & Ragwitz, M. (2018). Effectiveness and Efficiency of Auctions for Supporting Renewable Electricity—What Can We Learn from Recent Experiences? *Renewable Energy, 119*, 473–489.

World Energy Council. (2015). *World Energy Trilemma Priority Actions on Climate Change and How to Balance the Trilemma.* Available at: http://www.worldenergy.org/wp-content/uploads/2015/05/2015-World-Energy-Trilemma-Priority-actions-on-climate-change-and-how-to-balance-the-trilemma.pdf. Accessed 18 Feb 2016.

Zachmann, G. (2015). *When Will the EU Switch Away from Coal?* Available at: http://bruegel.org/2015/12/when-will-the-eu-switch-away-from-coal/. Accessed 28 Nov 2016.

CHAPTER 4

The Policy Stream

Abstract John Kingdon expresses the significance of ideas in his vision of the policy process by paraphrasing Victor Hugo: '*Greater than the tread of mighty armies is an idea whose time has come*'. The policy stream is where ideas are born and developed, combined and recombined, polished and prepared for their moment in the sun. This chapter follows the Brussels climate and energy policy community concerned as it trials tests and contests ideas about the 2030 targets in the several years leading up to 2014. The significant divisions within the policy community wrought by ideas are explored towards the end of the chapter.

Keywords Policy communities · Techno-economic modelling
Technology neutrality · European Union

4.1 INTRODUCTION

This chapter provides an account of the policy stream, the ideas that contributed to the debate about what might be an appropriate policy response. It is the second of three parallel chapters that describe each of the three streams of MSA. This chapter traces the development of the ideas and activities during the years and months leading up to the European Commission's proposal for a 2030 climate and energy framework and the role(s) played by the community of actors and coalitions,

© The Author(s) 2018   57
O. Fitch-Roy and J. Fairbrass, *Negotiating the EU's 2030
Climate and Energy Framework*, Progressive Energy Policy,
https://doi.org/10.1007/978-3-319-90948-6_4

both inside and outside the formal institutions of the EU, who were working closely on climate and energy policy.

Following a recap of the role of the policy stream in Sects. 4.2 and 4.3 presents a chronology of the status and presentation of various competing ideas about the appropriate policy response. It shows that while this area of policy is complex with many dimensions, the idea of 'technology neutrality' dominated the policy conflicts. The actors involved almost universally adopted a position either in favour of a technology-neutral 'single-target' approach or in favour of a technology-specific multiple-targets approach. Section 4.4 discusses the impact on the community of experts, the policy community, discussing the policy area of this polarisation, noting the challenges faced by the different groups' inability to communicate straightforwardly and the implications for the structure of the policy community of the process to reform the EU-ETS at the same time as negotiating the 2030 targets. Section 4.5 concludes the chapter.

4.2   MSA and the Policy Stream

The Multiple Streams Approach (MSA) proposes a number of factors that influence an idea's survival chances in the policy stream and gaining serious consideration (Kingdon 2010). Firstly, an idea or proposal needs to be technically feasible. Ideas that do not 'stack up' tend to be overlooked. Secondly, it needs to fit with the values of policymakers i.e. be 'acceptable'. It is also worth highlighting the point that the rate at which ideas are 'sifted' also depends on the characteristics, the size, degree of fragmentation and accessibility of the policy community itself (Kingdon 2010; Zahariadis 2014).

The following sections of this chapter now explore the ideas vying for acceptance in the policy stream during the agenda-setting phase that led to the 2030 targets, identifying those that proved especially significant. There is also an examination of the structure of the policy community.

4.3   The Problem Stream, 2009–2014

4.3.1   2009–2012: Analysis, Models and Roadmaps

Following the Copenhagen climate summit (COP 15) in late 2009 and the start of the second Barroso Commission in early 2010, the European Union's climate leadership may have been tarnished, but clear commitments had been set by the European Council to reduce greenhouse gas (GHG) emissions by 80–95% by 2050 (European Council 2009). A framework was in place for the period up to 2020 (although legislation on the energy efficiency component was still at an early stage[1]) but it was understood by policymakers that the Union needed a better grasp of the policy options and their technical and economic implications after 2020.

DG Energy, under Günter Oettinger's predecessor, Commissioner Andris Piebalgs had for some time been encouraging the energy industry, especially the electricity sector, to engage in some serious thinking about how the goal of a decarbonised energy system might be met (Interview 19, 2015; European Commission 2009). The electricity industry, represented by membership organisation, Eurelectric, responded by producing a detailed piece of scenario modelling entitled '*Power Choices: Pathways to Carbon-Neutral Electricity in Europe by 2050*'. The report presented a scenario for a decarbonised European energy system assuming that support for renewable energy is phased out by 2030 and in which the EU-ETS would be the primary tool (Eurelectric 2009).

At around the same time as Eurelectric produced 'Power Choices', the Commission, in informal discussions with a well-funded environmental organisation, the European Climate Foundation (ECF), expressed a wish for a '*trailblazing*' piece of analysis from civil society to set the terms of a discussion about options for a post-2020 climate and energy policy (Interview 11, 2015; Interview 19, 2015). Starting in August 2009, ECF formed a '*core reflection group*' of organisations including utilities, network companies, environmental non-governmental organisations (ENGOs) and manufacturers as well as a wider group of stakeholders with some access to the process including renewable energy firms and one oil major, namely, Shell (Interview 11, 2015).

The reflection group was invited to a series of workshops that informed a subsequent techno-economic modelling exercise (European Climate Foundation 2010a). The results of the modelling were presented in spring 2010 to newly appointed Commissioners Oettinger and Hedegaard as well as representatives from the other EU institutions in three reports under the title '*Roadmap 2050: A Practical Guide to a Prosperous, Low-Carbon Europe*' (European Climate Foundation 2010a).

---

[1]And there was already significant discussion, in the light of the ongoing economic crisis and recession dampening energy demand, to increase the 2020 energy efficiency target from 20 to 30%.

The work presented in the reports focused primarily on the electricity sector and assumed significant electrification of the heat and transport sectors in an almost completely decarbonised electricity system by the year 2050 (European Climate Foundation 2010b). In particular, ECF explored the implications of decarbonisation for electricity transmission.

The three 'no regrets' options identified by ECF's reports include:

1. Increasing renewable energy production;
2. Improving the energy efficiency of buildings; and
3. Strengthening internal market by improving Europe's cross-border energy interconnection (European Climate Foundation 2010b, c).

The policy recommendations for bringing about a decarbonised electricity system made in the reports include a strengthened EU-ETS cap and a new policy framework for climate and energy after the expiry of the 20/20/20 framework in 2020.

The ECF 'Roadmap 2050 project' and Eurelectric's 'Power Choices' reports made two important contributions to the debate. Firstly, they helped to draw attention to the lack of policy for keeping the EU on a trajectory towards its 2050 climate goals after 2020. Secondly, they initiated and set the terms of a wider conversation about what the post-2020 policy response ought to be and framed that conversation in terms of a 'roadmap' and 'no regrets' options such as renewable energy and energy efficiency.

The ECF modelling exercise, partly due to the collective responsibility taken by the reflection group was familiar to many people in the policy community. This familiarity, as well as the reports' general respectability, inspired some interest groups to make use of the analysis to support their own position. For example, the European Gas Advocacy Forum (EGF) explicitly drew on the ECF work to produce a report titled '*Making the Green Journey Work*' that underlined a role for gas as a fourth 'no regrets option' in several future scenarios (Interview 11; European Gas Advocacy Forum 2011). WWF later cited the ECF work in a report making a case for a 100% renewable energy future (WWF 2012). ECF acknowledged that in providing analysis to the community at large on an open basis, there was some risk of '*cherry picking numbers*' by groups with divergent interests, but it appears to have been a calculated risk. The decision to adopt such an open position was ultimately taken based on the assumption that the ECF report is not only original but demonstrates broad 'buy-in' and that policymakers in the Commission are 'wise enough' to detect 'self-interest' when they see it, as evidenced by the quotation that follows.

> *[Spin-off reports]* are secondary, and are always supported by certain influence groups. So the strength of the community-building report towards policymakers is always stronger than the secondary reports from a company or a group of companies that 'surprisingly' supports their product. (Interview 11, 2015)

In fact, there was so much scenario modelling being published that in preparing their own work in late 2011, the Commission produced an overview document as part of its impact assessment to review and compare the assumptions and results of various 'stakeholder scenarios' (European Commission 2011c).

There was also a level of interdependence between modelling exercises with studies using the outputs or underlying mathematical models of others (European Commission 2011c). The bulk of the modelling work undertaken by the policy community drew from two sources: the PRIMES model owned and operated by the Technical University of Athens on behalf of DG Energy and the International Energy Agency's (IEA) world energy model.

Each of the stakeholder groups publishing modelling studies came with its own set of policy recommendations which broadly seek to establish outcomes that reflect the interests of the publishers as 'no regrets' options. For example, ECF, broadly in favour of new energy targets, urged the Commission to *'consider the need for deployment targets beyond 2020 for key renewables generation technologies'* (European Climate Foundation 2010b, p. 4). Eurelectric's report places greater emphasis on *'a wide range of zero or very low carbon production options'* (Eurelectric 2009, p. 81) and the EGF pointed to economic savings in its scenario that are *'mainly a result of building fewer renewables'* (European Gas Advocacy Forum 2011, p. 24).

The scenario modelling 'frenzy' of 2009–2011, initiated at the request of the Commission, engaged parts of the policy community (i.e. ENGOs, energy producers and the Commission) in an intense discussion about ideas and suggestions for the post-2020 period. Notably, by demonstrating the ability to participate in detailed and meaningful analysis, it acted as a means of 'vetting' membership of the policy community.

In particular, ECF, a young organisation with little history of EU policy engagement rapidly established itself as a serious player in the climate and energy policy community through its analytical rigour. As one senior Commission official said in interview:

> "I think that, certainly the work that they *[ECF]* have done with their analysis has strengthened their position enormously, because it *[challenges]* the numbers that national member states or ourselves [the European Commission] produce in our own analyses"…"in terms of getting your point across, it is the analysis and the numbers that will be considerably more impactful than anything else". (Interview 10, 2015)

In 2011, the Commission produced two climate- and energy- related so-called 'roadmaps'. In March, it published '*A Roadmap for moving to a competitive low-carbon economy in 2050*', drafted by DG Climate Action setting out the challenge of economy-wide decarbonisation, confirming an 80% reduction in EU emissions by 2050 as an explicit policy goal rather than a purely political ambition, and establishing 2030 as the 'next' major milestone after 2020. A '*2050 Energy Roadmap*' drafted by DG Energy was published in December. The idea of a 40% GHG emissions reduction target was introduced and established as a main reference point for future discussion (European Commission 2011a, b). The roadmap concluded by declaring that the next logical step is to define a '2030 policy framework'.

4.3.2   2013: 'Targetology'

Equipped with a better understanding of the energy challenges and options Europe faced after 2020, and it being generally agreed that 2030 would act as a reasonable interim date on the way to 2050, the Commission then set the policy wheels in motion. In March 2013, it published a green paper on a '2030 framework for climate and energy policies' in which it was made clear that the Commission was thinking about proposing some targets for 2030 (European Commission 2013). Referring directly to the policy feedback from the 2020 package discussed in the previous chapter, the Commission set out the major questions about renewable energy, energy efficiency and an overarching climate target, asking the following

- Should there be multiple 2030 targets for GHG emission reductions and renewables, energy efficiency, etc. or just a single, GHG target?
- What should be the level of the target(s)?
- Should targets be legally binding? On individual member states?

The year 2013 then became a period of intense discussion and negotiation about what would be an appropriate level for and nature of the 2030 targets. The numerical rather than political focus of the debate led one observer to remark:

> Throughout this process, we've been worried about the conversation being based in the realm of 'targetology' and pure numbers. (Interview 2, 2015)

In 2013, the policy community was engaged in coalition building and position formulation, a process which caused some integration as well as some fracturing of the policy community. The following subsections describe actors' positions on five topics. First we look at the target for GHG emissions reduction, second the number of targets, third the level of a renewable energy target, fourth the level of an energy savings target and finally actors' positions in relation to European Integration.

The Target for GHG Emissions Reduction
One element of the framework that was never under any serious doubt was the inclusion of a GHG emissions reduction target for 2030, with its precise level becoming a major discussion point through 2013. Although a target for GHG emission reduction across the economy would also affect emissions not covered by the EU-ETS (in the so-called 'non-traded sectors'), it would necessarily have major impacts on the EU-ETS cap in the future. The modelling by DG Energy suggested that a figure of 40% was appropriate (European Commission 2011c), and that number was 'never seriously challenged' according to some Commission officials.

Since the impact assessment focused on scenarios for 35, 40 and 45% GHG emissions reductions some actors felt that 40% was presented as a credible 'middle way':

The Commission always proposes A, B, C on the presumption that that gives B the best chance. So ... that's why 35, 40 and 45 were run. It was to make 40 normative. (Interview 8, 2015)

Nevertheless, there were interests that sought to put forward a rationale for alternative levels. BUSINESSEUROPE, the energy-intensive industries and some member states such as the VISEGRAD group led by Poland lobbied consistently for a lower target or for no target at all, at least until after the Paris climate conference in 2015 (BUSINESSEUROPE 2013; IFIEC Europe 2013). The goal, if there must be a target, for many actors, including Energy Commissioner Oettinger, was 35% (Euractiv 2014h; European Commission 2014d; van Renssen 2014).

By contrast, within the EU, the United Kingdom (UK) had long been an advocate for ambitious action on climate change (Lorenzoni and Benson 2014). The UK's position on the GHG target, which was one of the first major national announcements with a personal blog by the British Energy Minister in May and an official speech in June 2013, was in line with the Commission's suggestion of 40%. There was also a stipulation that it should be increased to at least 50% should a very ambitious international agreement be achieved in Paris in December 2015 (Davey 2013a, b).

Much of the energy industry, both renewable producer groups and others such as the Magritte Group, was able to accept, overtly or tacitly, the idea of a 40% (or greater) GHG emission reduction target. Some of the interest groups, however, had to be 'fairly creative' to overcome objections by certain members. The secretariat and bulk of national electricity associations that make up Eurelectric, for example, were convinced by the argument put forward by the Commission and their own policy advisor to support a goal of 'at least 40%'. The Polish member, the coal-dominated Polski Komitet Energii Elektrycznej (PKEE), however, did not feel able to agree to such a target. To overcome this objection, the secretariat, in drafting a response to the 2030 green paper, included a footnote which stated that '*the Polish member association PKEE does not adhere to the views expressed in this paper, especially the positions on ETS*', an approach that was repeated in subsequent Eurelectric publications (Interview 8, 2015; Eurelectric 2013, p. 3; 2014, p. 1).

Some, including the Energy Commissioner, made a case that a 40% 2030 target compared to 1990 implied a rapid acceleration of effort, based on an assertion that, from 1990 to 2020, the expected reduction was just 20%, denoting a tripling of the annual reduction *rate* between 2020 and 2030.[2] The suggestion was that 'accelerating' the decarbonisation of the European economy after '*low-hanging* fruit' such as closure or displacement of much energy inefficient soviet-era Eastern European industry in the early 1990s and the ongoing economic slump as well as other '*one-time*' effects on carbon intensity have been accounted for is unrealistic (Interview 9, 2015; Euractiv 2014h). Although some in the oil and gas lobby may privately have supported the suggestion by Commissioner Oettinger that 40% was too ambitious, this was not a position that was aired publically (Statoil ASA 2013). The European coal sector had, for some time, been unique among the energy producer industries in its explicit resistance to ambitious climate policy, made this argument public by publishing a position paper late in the process which set out an argument that '*less climate ambition delivers more*' and proposed a 33% target (Euracoal 2014).

Number of Targets
A major source of difference within the policy community during 2013 was the disagreement over 'technology neutrality' and a 'targeted technology' policy. The idea of technology neutrality manifested itself in the form of a 'single-target' approach with only a GHG emissions reduction target, making the EU-ETS the primary European policy instrument for promoting the decarbonisation of the energy system and targeted technology policy by a 'multi-target' approach that included targets for renewable energy and energy efficiency.

The 'multiple targets' idea was most actively promoted by producer groups, particularly the renewable energy and energy efficiency supplier groups, who were seeking greater confidence in demand for their products in the future. The 'coalition of progressive energy companies' was set-up to argue in favour of multiple targets (Coalition of Progressive Energy Companies 2014) and its members, who were utilities with major renewables investments, all supported the idea (Interview 13, 2015; EDP 2013). In general, a central tenet of the argument in favour of multiple targets was that supporting new and emerging technologies could

---

[2]Although the change in the rate of reduction is far less abrupt if the reductions to 2020 assumed to have been achieved since 2009 when the 2020 targets were made law (still measured against the 1990 baseline).

help promote economic activity and create jobs that might bring possible energy security benefits via the greater use of domestic renewable energy.

One of the first organisations to acknowledge the importance of the single-/multiple-targets debate was the European Renewable Energy Council (EREC). EREC argued that the policy package should have wider aims than simply mitigating emissions, such as energy security and creation of new jobs in the renewables sectors (EREC 2013). In the College of Commissioners, Commissioners Hedegaard and Oettinger expressed lukewarm support for a multi-target framework while there was a clear split in the Commission services between DG Energy, which was pushing for multiple targets and DG Clima and the Secretariat General, which tended towards favouring a single-target approach (Interview 4, 2015; Interview 7, 2015).

Several groups were arguing in favour of technology neutrality and reinforcing the primacy of the EU-ETS in Europe's battle against climate change. There were two main arguments used to support this approach. The first was that because markets rather than policymakers determine things such as the energy mix, a firmly technology-neutral approach is inherently more economically efficient and therefore a less costly way of promoting the change needed to transform the European economy. One oil company lobbyist for stated that:

> ...what we weren't saying in this coalition wasn't that we need more gas, or we need more nuclear. We were saying that 'this policy [a single target] is the most cost-effective way to develop the 2030 [climate] target'. (Interview 25, 2015)

Single-target advocates also appealed to the concept of subsidiarity by stating that decisions about energy mixes are better at a national, rather than a European level.

The energy-intensive industries were unanimous that, if there must be climate ambition, it should be based on a single target (as well as aligned to an international climate agreement) (IFIEC Europe 2013; LaFarge 2013).

Partly due to an ideological commitment to the idea of technology neutrality and possibly due to its plans to build a fleet of new nuclear reactors, the UK was very strongly advocating a single-target approach, along with the Governments of Poland and the Czech Republic (Interview 17, 2015; Davey 2013b). The UK simultaneously deployed both the cost effectiveness *and* subsidiarity arguments:

...we will have to allow the Member States to make the right choices for themselves – especially the choices over which technologies to use in the low carbon transition......Countries should be free to pick the energy mix they prefer, and not be penalised for the choices they make – including on whether they choose nuclear to deliver their emissions reductions......And we will oppose a 2030 renewable energy target at an EU level as inflexible and unnecessary. (Davey 2013b)

This position put the UK at odds with most of the Green Growth Group, including Denmark and Germany which sought multiple targets (Government of Poland 2013; Government of the Czech Republic 2013; Government of Denmark 2013; Euractiv 2014d). The idea of a single target was supported by the gas industry (Eurogas 2013) and was central to the formation and activity of both the overlapping coalitions the Magritte Group and the Single-Target Coalition led by Shell, a company that had been making a case to the Commission for a single target since 2011 (Neslen 2015).

Meanwhile, Eurelectric again found itself internally conflicted about its position on one versus multiple targets. Some members, such as the Danish association, wanted to establish a multi-target position while a majority favoured a single-target approach (Interview 21, 2015). Partly because of this internal disagreement between members and partly because of the personal conviction of one of the group's public affairs advisors, Eurelectric side-stepped the topic to focus instead on issues such as the overall ambition of the package. Eurelectric's response to the Commission's green paper made no reference to whether a single or multi-target approach was preferred (Eurelectric 2013) and:

...we had to write very carefully fluffy language around the one target / three target space in order that they could co-sign things that they otherwise might have wanted to disagree about. So, some of our position papers at Eurelectric also involve some creatively vague wording at times. (Interview 8, 2015)

The inability of Eurelectric to agree on a strong position on the single-target issue was frustrating for some members for whom it was an important focus for lobbying. As a result, some pursued the issue in other coalitions such as the Magritte Group (Interview 1, 2015). The business-led Prince of Wales' Corporate Leaders Group (CLG) also equivocated on the issue of multiple versus single targets due to internal disagreement:

> Most of the EU CLG members would like to continue with a 20/20/20
> style framework supporting the GHG emissions target by setting binding
> targets for renewable energy and energy efficiency......However, this intro-
> duces the risk of weakening the carbon price and raising the cost of com-
> pliance if such policies are not correctly aligned. (The Prince of Wales's EU
> Corporate Leaders Group 2013)

A diverse coalition, Friends of ETS (FoETS), also took a position that
did not explicitly express a view on how many targets there ought to be
but did strike a de facto position in favour of a single target. In order to
accentuate the sense of crisis in the ETS, FoETS argued that the interaction between a GHG emission reduction target and other targets could
undermine the EU-ETS price. The campaign was unable to argue in
favour of multiple targets at the same time, especially since some of the
members were actively lobbying for a single target.

Target for Renewable Energy
One group with the clearest interest in the level of ambition of any
renewable energy target was the renewable energy producers. Therefore,
it was unsurprising that the umbrella group EREC moved to make a case
for a high target. As early as May 2011 it was arguing on behalf of the
renewables industries in favour of a renewable energy target of at least
45% by 2030 (EREC 2011). The Energy Roadmap 2050 published by
the same year, however, indicated that a level of 30% was commensurate
with achieving the 2050 decarbonisation goals (European Commission
2011b). In their responses to the Commission's consultation paper, the
solar energy industry group, EPIA continued to support EREC's 45%
figure while EWEA, EPIA's wind energy counterpart, did not, presaging
EWEA's departure from the group in January the following year, shortly
before its collapse (EWEA 2013; EPIA 2013; GSTEC 2014).

Target for Energy Savings
Although the level of energy savings was not discussed in the
Commission's green paper, the Coalition for Energy Savings com-
missioned a German research institute, Fraunhofer, to undertake a
study into the appropriate level of an energy efficiency target for 2030
(Eichhammer 2013). Not only did the work aim to show that the energy
savings targets need not interfere with emission reductions but that a
strong and binding target was the appropriate response. Fraunhofer's

modelling suggested that the optimum level for the energy savings target was 41%, which was below the baseline used by DG Ener in the 2050 roadmap modelling exercise, leading to the energy efficiency groups to suggest a 40% target (Eichhammer 2013; The Coalition for Energy Savings 2013). Meanwhile, the Commission, acknowledging the fact that energy efficiency policy area was in a state of flux with the 2012 Energy Efficiency Directive in the early stages of implementation and a review of progress towards the 2020 target imminent, did not suggest an appropriate level in the 2013 green paper.

More Versus Less Europe
The way that various actors framed their policy positions vis-à-vis European integration was significant. For example, the Magritte Group of utility CEOs, while highly critical of the approach taken to climate and energy policy before 2012, especially the 20/20/20 framework, pointed out the risk of policy developments such as the spread of national capacity remuneration mechanisms for rewarding electricity generation capacity 'fragmenting' the European energy market and policy landscape. These policy interventions were characterised by the Magritte group as a response to the surge in new variable renewable electricity generation resulting from the 20/20/20 framework. Crucially, the Magritte Group strongly advocated for a *European* rather than a *national* solution (Magritte Group 2013). Even though their rhetoric about European-level capacity remuneration weakened slightly, the Group's pro-European approach remained strong and was well received by the Commission. A close advisor to Commissioner Oettinger said:

> The Magritte Group…was saying that European energy policy was not working, which seemed to point the finger to Brussels as not having done what was necessary, but then they would say, at the same time that, in fact, what we needed was more Europeanisation. So, they were sort of critical of the state of play but saying it's because it's not enough, rather than because it's too much *[Europe]*. The message sounded, initially, very negative towards Brussels, the EU institutional set-up, but, as time progressed, it became clear that, in a way, we were allies on substance. (Interview 6, 2015)

Some politicians and officials in the Commission, Secretary General Catherine Day in particular, were felt to be more susceptible to arguments which either supported further European integration (or defended it from fragmentary forces) than purely economic arguments. Parts of the electricity industry wanted to emphasise the significance of the negotiations for European integration:

> "...if you think somebody is not interested in climate, not interested in energy, how do you explain that it matters to them?" "What we were saying to *[Secretary General, Catherine Day]* is, 'You should be interested in the energy-climate package because depending on how it plays out and depending on whether you've got multiple measures or multiple targets or different strengths of targets and so on, some of these are going to suit the internal energy market agenda better than others. You care about the internal energy markets agenda because it's part of the overall market harmonisation agenda, which is the DNA of the European project. And, if you get the 2030 package wrong, you have effectively let loose a virus in the DNA. Do you care about that?" ..."Therefore, one of the things that Eurelectric cared very much about was that whether there was one target or three targets, that they should be European level targets". (Interview 8, 2015)

4.3.3   2014: The Proposals

This section describes the months surrounding the Commission's publication of its proposals for the 2030 framework. First, we describe a growing consensus around the idea of a single target, which was punctured by the intervention of pro-multiple target member states. Second, we show that in the final weeks, the Commission perceived a stalemate between multi- and single-target proponents. Finally the reaction by various actors to the proposals are set out.

Consensus Builds Around a Single Target
Throughout 2013, the main point of coordination for the renewables industry, the EREC, had been diminishing in effectiveness and failed to follow-up its initially strong stance in favour of high and binding renewable energy targets or provide policymakers with straightforward channel of communication with the renewables sector. At the same time, the Magritte Group's strategy of delivering their message *en-masse* to the most senior national and European policymakers culminated in a meeting late in the year with Commission President Barroso (Interview 4, 2015).

By the end of the year, it was the Commission's view that there was considerable consensus around the idea of a single target, among member states as well as European civil society (Euractiv 2013a, b). The international oil company (IOC), BP, wrote to Energy Commissioner Oettinger in the summer making the case that, among other things, overly ambitious EU climate policy could push industries such as refining out of Europe (Neslen 2015, 2016).

Around the middle of 2013, DG Energy's energy efficiency unit was wrestling with whether it could possibly prepare its contribution in time for a January 2014 release, given a pre-determined commitment it had to review progress on the 2020 targets, a labour intensive task in itself. Eventually, the Commission decided that the energy efficiency part of the package would be announced according to a different timescale to the GHG emission target and the renewable energy target (Interview 28, 2016). The first instalment of the Commission's formal proposal for the 2030 package was set to be 'unveiled' on the 22nd of January 2014 and the year began with a flurry of activity from the parliament and member states.

Despite the weight of opinion that the framework was heading towards a single target, two events in early 2014 tempered expectations. First, over Christmas 2013, a joint letter from eight ministers including from Germany, France and Italy[3] was sent to the Commission requesting that 'robust' renewable energy targets be included in the package on grounds of creating '*more jobs and growth*' (Euractiv 2014a). The Prime Minister of Denmark also intervened by calling Commission President Barroso on the telephone the night before the Commission's decision in order to make the case for a renewable energy target (Euractiv 2014b). In addition to the overall effectiveness of the policy framework, a central argument in favour of strong targets on renewables was sustaining and creating jobs in the renewables and energy efficiency industries.

Secondly, the European Parliament also somewhat belatedly[4] entered the discussion with an 'own initiative' report produced on the 9th of January by the industry and environment committees and approved in plenary on the 27th. The report called for a 40% GHG emissions reduction target alongside 30% binding renewable energy and 40% energy

[3] As well as Austria, Belgium, Denmark, Ireland and Portugal.
[4] The report was, technically, a response to the 2013 proposals from the Commission.

efficiency targets, much to the relief of energy efficiency advocates who had invested significant effort lobbying Parliamentarians (Interview 10, 2015; Interview 12, 2015; Interview 17, 2015; Euractiv 2014h; European Parliament 2014; EuroACE 2014).

And so, despite the earlier expectations that the Commission would propose a single target, there were some media reports suggesting that, following the member state intervention lobbying for a renewables target, the Commission was set to introduce a non-binding renewable energy target some way below 30% (Euractiv 2014h).

Throughout the first half of 2014, advocates from all sides of the debate began addressing the Heads of State and Government of the European Council, which was due to meet in October to determine the final terms of the framework. Especially vocal were those that were unsettled by the perceived shift in favour of multiple targets that occurred over the winter. Rather than focussing on detailed policy, efforts tended to take a broader view on EU climate and energy policy. Some contributions reflected the shift from a technical to a political discussion by presenting contributions in the form of a 'manifesto' (IFIEC 2014; Eurelectric 2014).

It was not just interest groups in dispute about the fundamental structure of the framework. The package had become controversial within the Commission. Despite an uneasy agreement between Commissioners Hedegaard and Oettinger that some form of multiple-targets approach could be workable (although with ongoing disagreement about the level of the GHG target), some elements in the services, especially within DG Clima, were deeply committed to the idea of a single GHG target. This attachment to the idea was at least partially due to a pledge within the department to maintain the ETS as the 'cornerstone' of EU climate policy. After all, some of the people working on the policy had invested a decade of their professional life developing and tweaking the policy and therefore reluctant to see it usurped or undermined by further renewable energy and energy efficiency targets. Meanwhile, in DG Energy, many staff firmly believed that a multi-target approach which supported renewables and energy efficiency was more likely to prove effective than the ETS-only 'purist' approach. The dispute was ultimately resolved by Secretary General, Catherine Day when the drafting brief was taken from the two lead DGs, Clima and Ener and handed to a unit within the Secretariat General with the DGs relegated to simply reviewing 'bits and pieces of text' (Interview 7, 2015).

Stalemate

Lobbyists and policymakers both sensed that, during the autumn of 2013, a kind of stalemate had been reached. The disagreement among member states, civil society and even within the Commission caused the 2030 package to become what President Barroso described in a meeting of the College of Commissioners as particularly *'politically sensitive'* (European Commission 2014d, p. 23).

Ahead of the final text for the January 2014 Communication facing a vote in the College, Commissioners Hedegaard and Oettinger, Secretary General Day and President Barroso approved the content of the final draft (Euractiv 2014h). The outcome was designed to placate both sides of the debate and made use of 'wiggle room' afforded by the various options for the 'bindingness' of the targets: that is, by now, uncontroversial *at least* 40% GHG emissions target was to be coupled with a renewable energy target that was binding only at the European level. The level of the renewable energy target was determined on the basis of a yet-to-be published[5] analysis of policy impacts. The impact assessment projected that, if GHG emissions were reduced by 40% of 1990 levels by 2030, the consequential proportion of final energy consumption from renewable sources would be between 24 and 27%, depending on certain *'enabling conditions'* such as infrastructure improvements (European Commission 2014c; Euractiv 2014h).

The Commission selected at target from within this range, which Commissioner Oettinger explained to his colleagues in the College, was:

> a percentage that would almost automatically be engendered by the target of a 40% reduction in greenhouse gas emissions, according to the econometric models. (European Commission 2014d, p. 23)

On the 22nd of January and following a highly charged meeting of Commissioners lasting more than three hours, the Commission released the first of its proposals (Interview 8, 2015; European Commission 2014d). The Communication's *'A policy framework for climate and energy in the period from 2020 to 2030'* was published along with a detailed impact assessment, focussed primarily on the structure and level of the targets. It confirmed that the Commission was proposing a 40%

---

[5] The analysis was published as part of impact assessment alongside the Commission proposals on the 22nd of January.

GHG emissions reduction s target and a 27% renewable energy target binding at an EU level rather than a member state level. As expected, publication of the level and nature of the energy efficiency element of the proposed package was predicted later in 2014 with a level of 25% anticipated (European Commission 2014a).

The Response to the Initial Proposals

The initial, somewhat predictable, response from ENGOs was one of disappointment and an immediate call for more climate ambition and a renewable energy target that is binding on member states. Some Green MEPs and ENGOs also accused the Commission of moving too far to accommodate the demands of industry, especially the utilities of the Magritte Group (Interview 4, 2015; Euractiv 2014e, f). In defence of the proposals, Climate Commissioner Hedegaard wanted to maintain focus on the GHG emissions reduction target component of the package which she described as '*not a small thing, a big thing*' (Euractiv 2014f). In February, Greenpeace published a report singling out Europe's large utilities as the major barrier to effective climate and energy policy (Dallos 2014). The 'progressive' utilities such as EDP, DONG and SSE cautiously welcomed the proposals. However, the demise of the mortally wounded EREC as a coordination and contact point for the renewable energy industries was completed in March 2014 when the group was liquidated, leaving the renewable energy sector effectively locked out of the policy process for the remainder of the year (Interview 13, 2015; Euractiv 2014c).

Within a fortnight of the publication of the Commission's proposals in January, Energy Commissioner Oettinger publically aired his view that a 40% GHG emissions reduction target was '*too ambitious*' and '*unachievable*' to a receptive audience at a BUSINESSEUROPE event in Brussels, statements that appeared to be at odds with the will of the College of Commissioners (Euractiv 2014g, h).

The focus of the debate split in the first half of 2014. While all efforts were made to engage with the Commission on the forthcoming proposals on energy efficiency (with a review expected in the middle of the year), interaction with member states also increased in advance of the final decision by the Heads of State and Government at the European Council in October. Energy efficiency advocates felt that the period between January and July 2014 presented an opportunity to make their case for increasing the 25% savings hinted at in the January

Communication. Energy savings advocates attached energy efficiency's positive potential benefits for energy security to Russia's annexation of Crimea in March. Energy efficiency industry group, EuroACE, was also able to take advantage of an opportunity to present its ideas to an informal meeting of energy ministers in Athens in May convened specifically to discuss the topic of energy security (Joyce 2014; Greek Presidency of the Council of the European Union 2014).

Ahead of the publication of the Commission's proposal for the energy efficiency in the 2030 framework, the same group of member states that had written to President Barroso earlier in the year in support of a renewable energy target did so again, this time to make the case for an energy efficiency target (Garside 2014). The Commission's communication on energy efficiency in July suggested that a target in the region of 30% was supported by the evidence (European Commission 2014b).

Finally, in October 2014, the European Council broadly endorsed the Commission's proposals by creating a framework of an *at least* 40 GHG emissions reduction target, an *at least* 27% renewable energy target, binding at a European level and an *at least* 27% indicative energy efficiency target (European Council 2014).

4.4   A Fractured Policy Community

The previous sections portray the policy options debated and give an overview of the development of arguments and counter-arguments within the policy community between 2009 and 2014. This section now provides a brief description of the policy community, focussing on the key cleavages.

The actors with an interest in the 2030 targets between them held a wide range of often-incompatible views of the diagnosis of and solution to climate and energy policy problems. At the level of policy, there were three main groupings with strongly divergent positions:

- **Group 1**: Actors that resist all action on climate and/or energy from the EU. These include the energy-intensive industries such as steel and chemical manufacturers and are often organised at a European level by BUSINESSEUROPE although associations such as IFIEC are also important;
- **Group 2**: Actors that actively support action on climate change but perceive some types of policy (such as renewable energy targets) as

inappropriate or threatening in some way. This group includes the regulated electricity sector and some environmentally minded businesses such as some members of the CLG;

- **Group 3**: Actors that either feel that technology-specific policy should be pursued with utmost urgency, such as ENGOs like Friends of the Earth and Greenpeace or those that have a commercial economic interest in certain types of action such as the renewable energy producer groups and energy efficiency equipment industry.

Groups 2 and 3 together made up what can be described as an 'epistemic community' or '*network of professionals with recognised expertise and competence....and an authoritative claim to policy-relevant knowledge... [with a] shared set of normative and principled beliefs*' (Haas 1992, p. 3; Dunlop 2000). While not homogenous in policy positions and opinions, these actors shared the basic underlying premise of the need to tackle climate change and there was widespread and frequent interaction between its members, not least through mutual critiques of modelling and other analytical interventions. This community included a large proportion of the Commission officials working on drafting the proposals (Interview 8, 2015). Group 1, on the other hand, had an equally strongly shared worldview but did not interact strongly with the other groups. The energy-intensive industry, in particular, was both perceived, and perceived itself to be, outside of the core of agenda-shaping actors. While accusations of 'climate denialism' are made about the group 1 actors by actors from groups 2 and 3, a great deal of care is taken not to engage in disputes about the validity of climate science (see Euracoal 2015, for instance).

There could also be said to exist a fourth group which, while sharing many of the beliefs and principles of group 2, saw a strategic opportunity to couple the campaigning strategies at which group 3 was especially adept with the political credibility of some of the actors in group 2, especially large businesses. Figure 4.1 shows the structure of the policy community and the relationship of the four distinct groups of actors.

The ongoing EU-ETS reform process drove this structuring of the climate-conscious portion of the policy community. In order to secure reform of the EU-ETS, a calculation was made that cast aside some of the politically challenging topics of renewable energy and energy

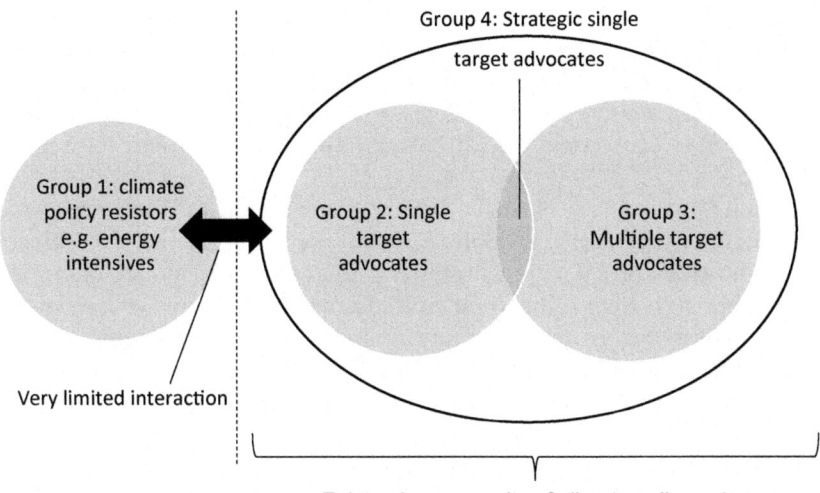

Fig. 4.1 Structure of the policy community in Brussels

efficiency to produce a louder chorus in favour of stronger overall climate action. Although rooted in the policy debate about ETS reform, the implications of this structuring for the 2030 targets debate were significant. We discuss this phenomenon and its role in more detail in Chapter 6 but here we argue that the primary effect was to subdue interest groups' ability to argue in favour of multiple targets while actively framing the ETS as the primary issue in EU climate and energy policy.

The general agreement within each group about the nature and importance of problems and about the type of solutions that might be appropriate led to a high degree of interaction between each of the two sides of the debate and policymakers with corresponding views. One lobbyist went as far as to claim that:

> ...from my point of view as a lobbyist, there were people in the Commission who I regarded as entirely part of the team I was working with and for in this process......and, I mean, there is no question that those who were arguing *[the other side of the debate]* were having those conversations with their Commission allies.

This closeness of relationships and shared beliefs among external stakeholders and between stakeholders and Commission staff rests on two elements. First, there were the interpersonal relationships between individuals '*because we've known each other for years and we're working on the same subjects*' (Interview 26, 2016) and '*it's all about trust*' (Interview 30, 2016).

Second were the mutual resource dependencies. The resources exchanged were highly variable. Clearly, the attention of Commission staff who draft or are able to influence the content of policy document is valuable in its own right to outside stakeholders who provide resources such as information in return. Information may be a critique, political intelligence or new analysis but, as one lobbyist put it, the aim of interest groups is to become '*...the captains of information*' (Interview 30, 2016).

Resource exchange often forms the basis of informal or ad hoc coalitions of interest groups (Baumgartner et al. 2009). In groups such as the Single-Target Coalition, FoETS and the Magritte Group, as well as a perceived legitimacy (a resource of primary importance), members may bring analytical skills, communications or PR skills, contacts, NGO-style campaigning skills, which are especially sought after by large companies which are '*...just not set-up to operate like that*' (Interview 1, 2015). In some cases, especially the FoETS coalition, there was a clear adaptation of policy preferences (in favour of a single target, for example) in order to enable stronger, more resilient coalitions to be built. This kind of flexible, dynamic coalition building built on shared interests, compromise, trust and resource exchange is far more prevalent in the self-appointed 'progressive' energy community than in the traditional industries that tend toward more hierarchical, inflexible modes of collective action. There was, however, a surprising lack of coordination between the renewables sector and the energy efficiency sector, which had a clear common interest in building a case for multiple targets. This could have been due to the lack of coordination within the renewables sector itself, which struggled even to overcome its own technology divisions (Interview 13, 2015).

The fragmentation of the policy community led to a great deal of suspicion between the two halves. A fragmented policy community tends to result in disjointed policy or '*the right hand not knowing what the left hand is doing*' as Kingdon (2010, p. 118) puts it. It certainly appears to have been the case in climate and energy policy that two halves of the policy community had very different perspectives on the policy process. While heavy industry was doing its best to raise the profile of so-called carbon leakage problem and the deindustrialisation of Europe, the topic rarely featured in discussions among the wider community, leading to a sense of frustration of energy-intensive industry lobbyists some of whom were suspicious of policymakers' motives and attitudes:

> ...of course, because *[the Secretary General]* has supported Clima blindly and therefore... You know, just in general. DG Clima have the role where they completely ignore industry...she blindly backed Clima, yeah, blindly. On everything. So the people who had anything trying to say... 'Well, yeah the impact on industry might be bad.' Basically, you know, she *[gave]* a little bit here, little bit there, but on the big issues we were completely ignored. Or I would say actively ignored. You can say that. (Interview 20, 2015)

It is also common in the traditional energy industries to discuss the funding arrangements of the ENGOs in pejorative terms, many of whom receive grant funding from the Commission as a method for 'buying' support as reflected in the quotation that follows:

> "So what the Commission has done is created an echo chamber in Brussels by funding people who are supportive of the policy direction that they wish to go in. You can have different views on that."... "I don't think it's a very good system of government. It's conflicted. If there are paid lobbyists in town and the lobbyists are paid by the organisation that they are lobbying, then, of course, the institutions will hear exactly what they've paid to hear in the same way that my message is exactly what my industry members want me to say because I'm paid to say that." (Interview 27, 2016)
>
> "We have to pay for everything. And it's very expensive." "[But] basically DG Clima[6] is sponsoring all the NGOs. To make their cause. That's the real lobby." Without the Commission, that lobby would not exist." (Interview 20, 2015)

The lack of common agreement about the relative salience of particular problems and a mistrust between the two camps is evident in a report produced by the coal producers' association designed to parody the

---

[6] The LIFE+ programmes that the interviewee was referring to are actually funded by DG Environment.

style and content of reports from ENGOs, the latter being described by Euroacoal as '*dubious*' (Interview 27, 2016).

4.5   Conclusion

MSA proposes that the policy stream represents a primaeval soup in which policy ideas compete for acceptance within a policy community (Kingdon 2010). This chapter examined the ideas that had currency in the debate about the 2030 targets and charted their fates through time during the agenda-setting process.

The 2030 framework proposed by the Commission and endorsed by the European Council contained multiple targets. The renewable energy and energy efficiency targets were, however, lower than advocates had hoped for and they were not directly binding on the member states, probably reducing their impact. Some saw this as a desirable outcome. In the words of one oil and gas lobbyist: '*the Commission probably got it right because nobody was happy. It smells like a good compromise*' (Interview 9, 2015).

Following a period of techno-economic modelling between 2011 and 2013 by various actors, several key contested ideas emerged about what the 2030 climate and energy package should contain. In particular, the idea of single, technology-neutral GHG-only target was a key battleground, defining actors' position relative to the 2030 targets and was of fundamental importance to the structure of the policy community. The self-identified 'pro environmental' groupings of the policy community split into single and multiple-target camps. In general, businesses supported a single target and ENGOs, along with the renewable energy and energy efficiency industries, multiple targets. However, the negotiating of lobbying coalitions had an impact on these preferences.

By the end of 2013, the idea of a single, GHG-only target, technology-neutral target dominated the policymaking agenda, with the Commission and others perceiving a consensus around the idea. Intervention by several member states prompted the creation of a compromise multiple-target package.

The single-target idea proved to be important for several reasons. Business actors such as the Magritte Group expended great effort to argue that a single target was both more market-based *and* more 'European' than multiple targets and therefore in line with the values of leading European politicians. At the same time, environmental coalitions such as 'Friends of ETS' were formed that adopted a de facto single-target position in spite of the natural scepticism for the idea held by many in the renewable energy and ENGO communities.

This chapter has described the policy stream and the community in which the discussion took place. MSA proposes that the overarching political context influences the convergence of problems and policies; the next chapter presents the politics stream.

REFERENCES

Baumgartner, F. R., et al. (2009). *Lobbying and Policy Change: Who Wins, Who Loses, and Why*. Chicago, IL: University of Chicago Press.

BUSINESSEUROPE. (2013). *2030 Green Paper Response*. Available at: https://www.businesseurope.eu/sites/buseur/files/media/imported/2013-00699-E.pdf. Accessed 19 Feb 2016.

Coalition of Progressive Energy Companies. (2014). *Energy Companies Call for an Ambitious and Binding Renewables Target for 2030*. Available at: http://www.eneco.com/nl/~/media/cor/pdf/organisatie/030314coalitionletterrenewablestarget2030.ashx. Accessed 4 Apr 2015.

Dallos, G. (2014). *Locked in the Past: Why Europe's Big Energy Companies Fear Change*. Hamburg: Greenpeace.

Davey, E. (2013a). *Department of Energy and Climate Change Blog: Europe Must Stay Ambitious on Climate Change*. Available at: http://blog.decc.gov.uk/2013/05/28/europe-must-stay-ambitious-on-climate-change/. Accessed 4 Nov 2014.

Davey, E. (2013b). *Edward Davey Speech: Ambitious and Flexible—Europe's 2030 Framework for Emissions Reduction*. Available at: https://www.gov.uk/government/speeches/edward-davey-speech-ambitious-and-flexible-europes-2030-framework-for-emissions-reduction. Accessed 4 Nov 2014.

Dunlop, C. (2000). Epistemic Communities: A Reply to Toke. *Politics, 20*(3), 137–144.

EDP. (2013). *2030 Green Paper Response*. Available at: https://crowdsourcing.simpolproject.eu/static/staticdata/gpc/consultations/edp_energias_de_portugal.pdf. Accessed 15 Apr 2016.

Eichhammer, W. (2013). *Analysis of a European Reference Target System for 2030. Energy Savings 2030: On the 2050 Pathway*. Available at: http://www.isi.fraunhofer.de/isi-wAssets/docs/x/de/publikationen/Fraunhofer-ISI_ReferenceTargetSystemReport.pdf. Accessed 11 Feb 2016.

EPIA. (2013). *2030 Green Paper Response*. Available at: https://crowdsourcing.simpolproject.eu/static/staticdata/gpc/consultations/epia.pdf. Accessed 18 Apr 2016.

EREC. (2011). *45% by 2030: Towards a Truly Sustainable Energy System in the EU.* Brussels.

EREC. (2013). *Hat-Trick 2030, Renewable Energy, Energy Efficiency, Greenhouse Gas.* Brussels.

Euracoal. (2014). *Why Less Climate Ambition Would Deliver More for the EU.* Brussels.

Euracoal. (2015). *Climate Change.* euracoal.eu. Available at: https://euracoal. eu/coal/climate-change/. Accessed 12 Oct 2016.

Euractiv. (2013a). *Hedegaard: More 2030 Climate Targets Would Be 'Wise'.* Euractiv.com. Available at: http://www.euractiv.com/energy/hedegaard-2030-climate-targets-w-news-530979. Accessed 19 Apr 2016.

Euractiv. (2013b). *Oettinger Hails 'Wide Agreement' on 2030 Energy Targets, but Doubts Persist.* Euractiv.com. Available at: http://www.euractiv.com/energy/2030-energy-target-doubts-oettin-news-530613. Accessed 4 May 2016.

Euractiv. (2014a). *Big EU Guns Fire for 'Crucial' 2030 Renewable Targets.* Euractiv.com. Available at: http://www.euractiv.com/energy/big-eu-guns-fire-crucial-2030-re-news-532608. Accessed 4 May 2016.

Euractiv. (2014b). *Denmark Signals Fight for Tougher 2030 Climate and Clean Energy Goals.* Euractive.com. Available at: http://www.euractiv.com/energy/denmark-signals-fight-tougher-20-news-533025. Accessed 2 Feb 2016.

Euractiv. (2014c). *EU Sets Out 'Walk Now, Sprint Later' 2030 Clean Energy Vision.* Euractiv.com. Available at: http://www.euractiv.com/energy/eu-sets-walk-sprint-2030-clean-e-news-532960. Accessed 4 May 2016.

Euractiv. (2014d). *Germany Calls for Three 2030 Climate and Energy Targets.* Available at: http://www.euractiv.com/section/energy/news/germany-calls-for-three-2030-climate-and-energy-targets/. Accessed 4 May 2016.

Euractiv. (2014e). *Green MEP: Lobbyists Stopped Ambitious EU Energy Targets.* Euractiv.com. Available at: http://www.euractiv.com/sections/energy/green-mep-lobbyists-stopped-ambitious-eu-energy-targets-309867. Accessed 23 Feb 2015.

Euractiv. (2014f). *Green MEPs, NGOs Protest Commission's New 2030 Climate and Energy Targets.* Available at: http://www.euractiv.com/section/energy/video/green-meps-ngos-protest-commission-s-new-2030-climate-and-energy-targets/. Accessed 2 Apr 2016.

Euractiv. (2014g). *Oettinger Feels the Heat Over Climate Remarks.* Euractiv.com. Available at: https://www.euractiv.com/section/energy/news/oettinger-feels-the-heat-over-climate-remarks/. Accessed 5 Apr 2016.

Euractiv. (2014h). *Oettinger Rallies Opposition to 2030 CO2 Target.* Available at: https://www.euractiv.com/section/trade-society/news/oettinger-rallies-opposition-to-2030-co2-target/. Accessed 5 Apr 2016.

Eurelectric. (2009). *Power Choices: Pathways to Carbon-Neutral Electricity in Europe by 2050.* Available at: www.eurelectric.org/PowerChoices2050/. Accessed 4 Apr 2016.

Eurelectric. (2013). *2030 Green Paper Response.* Available at: https://www3. eurelectric.org/media/110882/eurelectric_-_2030_green_paper_consultation_response_-_final-2013-030-0486-01-e.pdf. Accessed 13 Apr 2016.

Eurelectric. (2014). *Power for a Competitive Europe: A Manifesto for a Balance, More Efficient European Energy Policy.* Available at: http://www.eurelectric. org/media/119468/manifesto_designed-2014-030-0083-01-e.pdf. Accessed 21 Apr 2016.

EuroACE. (2014). *EuroACE Congratulates European Parliament for Its Economic Rationale in Calling for a Binding 40% Energy Efficiency Target and Sectoral Target for Buildings.* Brussels.

Eurogas. (2013). *The Eurogas 10 Point Plan to 2030.* Available at: http://www. eurogas.org/uploads/media/The_Eurogas_10-Point_Plan_to_2030.pdf. Accessed 14 July 2015.

European Climate Foundation. (2010a). *Newsletter Summer 2010.* Available at: http://www.europeanclimate.org/documents/ECF_Newsletter_Summer 2010.pdf. Accessed 31 Mar 2016.

European Climate Foundation. (2010b). *Roadmap 2050: A Practical Guide to a Prosperous, Low-Carbon Europe* (Vol. 2). Available at: http://www. roadmap2050.eu/. Accessed 31 Mar 2016.

European Climate Foundation. (2010c). *Roadmap 2050: A Practical Guide to a Prosperous, Low-Carbon Europe* (Vol.1). Available at: http://www. roadmap2050.eu/. Accessed 31 Mar 2016.

European Commission. (2009). *Commissioner Piebalgs Welcomes the Commitment of European Electricity Companies to Achieve a Carbon-Neutral Power Supply by 2050.* Available at: http://europa.eu/rapid/press-release_IP-09-417_en.pdf. Accessed 31 Mar 2016.

European Commission. (2011a). *A Roadmap for Moving to a Competitive Low Carbon Economy in 2050.* Available at: http://ec.europa.eu/clima/documentation/roadmap/docs/com_2011_112_en.pdf. Accessed 4 Apr 2016.

European Commission. (2011b). *Energy Roadmap 2050.* Available at: https://ec.europa.eu/energy/sites/ener/files/documents/2012_energy_ roadmap_2050_en_0.pdf. Accessed 14 July 2015.

European Commission. (2011c). *Energy Roadmap 2050 Impact Assessment, Part 2/2: Accompanying the Document Energy Roadmap 2050.* Available at: https://ec.europa.eu/energy/sites/ener/files/documents/sec_2011_1565_ part2.pdf. Accessed 4 April 2016.

European Commission. (2013). *Green Paper: A 2030 Framework for Climate and Energy Policies.* Available at: http://ec.europa.eu/transparency/regdoc/ rep/1/2013/EN/1-2013-169-EN-F1-1.pdf. Accessed 10 Feb 2014.

European Commission. (2014a). *A Policy Framework for Climate and Energy in the Period from 2020 to 2030.* Available at: http://eur-lex.europa.eu/ legal-content/EN/TXT/PDF/?uri=CELEX:52014DC0015&from=EN. Accessed 4 Feb 2014.

European Commission. (2014b). *Energy Efficiency and Its Contribution to Energy Security and the 2030 Framework for Climate and Energy Policy.* Available at: http://eur-lex.europa.eu/legal-content/EN/TXT/?uri=CELEX:52014DC0520. Accessed 20 Apr 2016.

European Commission. (2014c). *Impact Assessment—A Policy Framework for Climate and Energy in the Period from 2020 Up to 2030.* Available at: http://ec.europa.eu/smart-regulation/impact/ia_carried_out/docs/ia_2014/swd_2014_0015_en.pdf. Accessed 13 July 2015.

European Commission. (2014d). *Minutes of the 2072nd Meeting of the Commission Held in Brussels (Berlaymont) on Wednesday 22 January.* Available at: http://ec.europa.eu/transparency/regdoc/rep/10061/2014/EN/10061-2014-2072-EN-F1-1.Pdf. Accessed 20 Apr 2016.

European Council. (2009). *29/30 October 2009: Conclusions.* Available at: http://www.consilium.europa.eu/uedocs/cms_data/docs/pressdata/en/ec/110889.pdf. Accessed 31 March 2016.

European Council. (2014). *European Council (23 and 24 October 2014) Conclusions on 2030 Climate and Energy Policy Framework.* Available at: http://www.consilium.europa.eu/uedocs/cms_data/docs/pressdata/en/ec/145397.pdf. Accessed 24 Oct 2014.

European Gas Advocacy Forum. (2011). *Making the Green Journey Work.* Brussels.

European Parliament. (2014). *Report on a 2030 Framework for Climate and Energy Policies.* Available at: http://www.europarl.europa.eu/sides/getDoc.do?pubRef=-//EP//NONSGML+REPORT+A7-2014-0047+0+DOC+PDF+V0//EN. Accessed 31 Jan 2016.

EWEA. (2013). *2030 Green Paper Response.* Available at: https://crowdsourcing.simpolproject.eu/static/staticdata/gpc/consultations/ewea.pdf. Accessed 18 Apr 2016.

Garside, B. (2014). *Ministers from 7 EU Nations Call for Binding Energy Saving Goal.* Reuters.com. Available at: http://uk.reuters.com/article/eu-energy-efficiency-idUKL5N0OY4KQ20140617. Accessed 22 Apr 2016.

Government of Denmark. (2013). *2030 Green Paper Response.* Copenhagen.

Government of Poland. (2013). *2030 Green Paper Response.* Brussels.

Government of the Czech Republic. (2013). *2030 Green Paper Response.* Brussels.

Greek Presidency of the Council of the European Union. (2014). *Informal Meeting of Energy Ministers Athens, 15–16 May* ('Energy Security' Discussion Paper). Available at: http://gr2014.eu/sites/default/files/DiscussionPaperonEnergySecurity.pdf. Accessed 27 Apr 2016.

GSTEC. (2014). *Belgium: European Renewable Energy Council (EREC) Is History.* Available at: http://www.gstec.org/content/belgium-european-renewable-energy-council-erec-history. Accessed 15 Mar 2016.

Haas, P. M. (1992). Epistemic Communities and International Policy Coordination. *International Organization, 46*(1), 1–35.

IFIEC. (2014). *Manifesto: Europe's Manufacturing Industry CEOs Call Upon Heads of State to Streamline 2030 Strategy Towards Growth and Jobs*. Brussels.

IFIEC Europe. (2013). *2030 Green Paper Response*. Brussels.

Joyce, A. (2014). *Speech to the Informal Energy Ministers Meetings—Athens—16 May*. Athens: Financing of Energy Efficiency Measures.

Kingdon, J. W. (2010). *Agendas, Alternatives, and Public Policies* (2nd ed.). Harlow: Pearson.

LaFarge. (2013). *2030 Green Paper Response*. Brussels.

Lorenzoni, I., & Benson, D. (2014). Radical Institutional Change in Environmental Governance: Explaining the Origins of the UK Climate Change Act 2008 Through Discursive and Streams Perspectives. *Global Environmental Change, 29*, 10–21.

Magritte Group. (2013). *Press Release: Heads of 12 Leading European Energy Companies Propose Concrete Measures to Rebuild Europe's Energy Policy*. Available at: https://www.engie.com/wp-content/uploads/2013/11/12CEO_VA_v4. pdf. Accessed 19 Feb 2014.

Neslen, A. (2015). *Shell Lobbied to Undermine EU Renewables Targets, Documents Reveal*. Available at: http://www.theguardian.com/environ-ment/2015/apr/27/shell-lobbied-to-undermine-eu-renewables-targets-doc-uments-reveal. Accessed 27 Apr 2015.

Neslen, A. (2016). *EU Dropped Climate Policies After BP Threat of Oil Industry 'Exodus'*. Available at: http://www.theguardian.com/environment/2016/apr/20/eu-dropped-climate-policies-after-bp-threat-oil-industry-exodus. Accessed 22 Apr 2016.

van Renssen, S. (2014). *Climate Policy Bumps into Competitiveness in Europe*. Energypost.eu. Available at: http://www.energypost.eu/climate-policy-bumps-competitiveness-europe/. Accessed 13 July 2016.

Statoil ASA. (2013). *2030 Green Paper Response*. Brussels.

The Coalition for Energy Savings. (2013). *A Binding Energy Savings Target for 2030: The Cornerstone for Mutually Supporting Climate and Energy Policies*. Available at: https://www.eurima.org/uploads/ModuleXtender/Publications/105/20131011_Coalition_position_on_2030.pdf. Accessed 7 Apr 2016.

The Prince of Wales's EU Corporate Leaders Group. (2013). *2030 Green Paper Response*. Cambridge: University of Cambridge.

WWF. (2012). *Re-energising Europe, Cutting Energy Related Emissions the Right Way*. Available at: http://awsassets.panda.org/downloads/cutting_energy_related_emissions_the_right_way_.pdf. Accessed 5 April 2016.

Zahariadis, N. (2014). Ambiguity and Multiple Streams. In P. A. Sabatier & C. M. Weible (Eds.), *Theories of the Policy Process* (pp. 25–57). Boulder, CO: Westview Press.

CHAPTER 5

The Politics Stream

Abstract The politics stream represents the large-scale political trends in which the policy process is embedded. This chapter provides an account of the politics stream, tracing important national and European political trends such as the rise of populist sentiment, as reflected in the 2014 European Parliamentary elections and the decline of public concern for climate issues following the 2009 international climate conference. The member state positions ahead of the October 2014 EU summit which decided the 2030 targets are analysed, concluding that consensus around climate and energy priorities was in short supply.

Keywords European Parliament · Public opinion · European Council Climate change

5.1 Introduction

In Multiple Stream Approach (MSA), the politics stream is said to flow independently of the community of specialists concerned with policy problems and solutions (Kingdon 2010). In contrast to a wider definition of 'politics' that is utilised by much political science scholarship and indeed much of this book, 'politics' in the sense used in MSA is ascribed a fairly narrow definition based on electoral, partisan or other issues which one might find discussed in the 'politics' section of a national

© The Author(s) 2018   87
O. Fitch-Roy and J. Fairbrass, *Negotiating the EU's 2030 Climate and Energy Framework*, Progressive Energy Policy,
https://doi.org/10.1007/978-3-319-90948-6_5

newspaper (Kingdon 2010, p. 145; Zahariadis 2007). Far from being exogenous to the policy process, MSA sees the political context as a crucial '*promoter or inhibiter*' of an idea's progress on the policymaking agenda (Kingdon 2010, p. 163).

This chapter shows that European public attitudes to both the EU and to climate change were souring during the agenda-setting phase of the 2030 climate and energy framework. It also reveals that among member states, despite a number of alliances and coalitions, the situation is best described as 'confused' with little consensus across all aspects of the policy. Additionally, the chapter provides a brief overview of national positions adopted at the October 2014 European Council summit, the critical meeting that decided the targets.

Following an overview of the role of the politics stream in MSA, Sect. 5.3 describes the prevailing European 'political mood', including a rise in Euroscepticism and a downgrading of the status of the challenge of climate change between 2009 and 2013. Section 5.4 surveys the political dynamics in the European Council ahead of the October summit. Section 5.5 concludes the chapter.

5.2   MSA and the Politics Stream

MSA proposes that the political context in which policy problems are discussed has implications for the coupling of problems and policies and the emerging policymaking agenda (Kingdon 2010). In essence, the politics stream '*reveals the extent of consensus or dissent in the broader political arena on specific issues*' (Herweg and Zahariadis 2018).

In a national or federal context, MSA tends to include three main elements within the politics stream: the national mood, pressure group campaigns and administrative or legislative turnover. The unique and somewhat idiosyncratic nature of the EU policy process means that employing the MSA in an EU context requires some modification to the approach. While a range adaptations are proposed by various authors (see Herweg and Zahariadis 2018 for an exhaustive review), Zahariadis (2008, p. 518) proposes the factors that make up the EU politics stream that meet the needs of this study as: '*the balance of Council member national and partisan affiliation, the ideological balance of parties in Parliament, and the European mood*' (Zahariadis 2008, p. 518). Accordingly, to structure this chapter, we focus on several distinct factors:

- The European mood and 2014 European Parliamentary elections;
- National positions taken at the October 2014 European Council.

5.3   The European Mood and 2014 Parliamentary Elections

Direct engagement with a 'European public' by EU politicians is undoubtedly weaker than in national settings. Moreover, the concept of an *'integrated public sphere'* in the EU certainly has its critics who point to European politicians' lack of direct accountability to voters, taking it as an indication that it may not exist (Princen 2007). Nevertheless, on certain issues, a 'European mood' or *'climate of the times'* can be detected and described which is strongest when the decision-making in question tends to take place at the European level (Van de Steeg 2006). This includes debates concerning the nature and future of the EU and Europe-wide or global issues such as an economic crisis or climate change.

Below, we discuss the European mood in terms of two topics that reflect issues that are generally transnational in nature and more specifically relevant to EU climate and energy policymaking. These are first public opinion about, and trust in, the EU and its institutions and, second, European citizens' views about the threat that climate change poses to Europe and the world. We also briefly discuss the results of the 2014 parliamentary elections in which populist and anti-EU parties increased their representation, marking an important political trend.

5.3.1   Public Opinion About the EU

For many years, EU scholars assumed that European citizens tended to be largely disinterested in European integration, and would be unlikely to either oppose or resist it. This assessment is based on the contention that EU integration was an 'élite business', and studied as such. This so-called 'permissive consensus' among citizens implied a generalised but shallow-rooted acceptance of the EU politics, leaving élites free to continue the process of integration largely unfettered by public opinion (Lindberg and Scheingold 1970). However, it could be argued that the permissive consensus subsided under the weight of increased public scrutiny following the Maastricht treaty in the early 1990s, to be replaced with what some scholars term a *'constraining dissensus'* among European citizens. Arguably, this was particularly evident following the testing period in which an EU constitution was rejected by French and Dutch voters in referendums in the mid-2000s (Marks and Hooghe 2009, p. 5; Hurrelmann 2007). By 2013, despite that year being labelled the *'Year of the European Citizens'* (European Commission 2013b), it was clear that trust in the EU, which had been steadily eroding for some time, was becoming a prominent political issue.

The evolution of a global financial crisis into the Eurozone sovereign debt crisis in 2009 divided Europe into 'creditor' countries such as Germany which had, before the crisis been running current account surpluses and 'debtor' countries such as Greece which, following years of entrenched current account deficits, risked default and widespread financial turmoil (Lane 2012). The latter group had been 'bailed out' with loans from other EU members and international organisations. Voters in both sets of countries were unhappy. Those in creditor countries such as Germany felt that they had been forced into a situation where they had to prop up the public finances of debtor countries 'ruined' by irresponsible financial practices, while the public opinion in debtor countries was inflamed by the tough and sometimes disciplinarian 'austerity' fiscal policies enforced from outside, largely by the EU (Lapavitsas et al. 2010; Lane 2012; Crum 2013). The Eurozone crisis contributed to public scepticism about the EU, so-called Euroscepticism, once considered a peculiarly British malaise, becoming a Europe-wide phenomenon with sharp declines in public trust measured in nearly all EU countries (Torreblanca and Leonard 2013).

This growing antipathy towards the EU was manifest in the growing importance of populist, anti-EU parties from both the left and right of the political spectrum. In the 2014 European elections, strongly anti-EU parties won around 100 of the 751 seats in the European Parliament with almost a third of MEPs from broadly anti-establishment parties (*The Economist* 2014; Spiegel and Carnegy 2014). The shift in the balance of power in the Parliament itself, however, was fairly modest with the European People's Party (EPP) and the Socialist and Democratic Alliance (S&D) and the Alliance of Liberals and Democrats (ALDE), all pro-EU groupings, maintaining a combined majority, supported by informal voting agreements on key legislation (Bressanelli et al. 2016; VoteWatch Europe 2015). Overall, the results of the election were symptomatic of growing scepticism across Europe about the EU's ability and authority to deal with the problems facing citizens.

5.3.2 European Opinion About Climate Change

It was not only European policymakers who were left feeling deflated in the years following the 2009 UN climate conference in Copenhagen. EU citizens' perception of climate change as a serious global problem had also been waning. In the summer of 2011, climate change was still seen as important by respondents to EU surveys, ranking second only to poverty, lack of food and drinking water 'as the most serious problem facing the world as a whole'[1] (European Commission 2011). However, by the end of 2013, climate change had been overtaken by 'the economic situation' and by 2015 had been pushed into fourth place among citizens' concerns by 'international terrorism' (European Commission 2014, 2015b). The proportion of respondents to the EU's Eurobarometer survey of European attitudes and opinion believing climate change to be the 'single most serious problem facing the world' fell from 20% in June 2011 to 16% in December 2013 and by 2015 it had fallen further to 15% (European Commission 2008, 2009, 2011, 2014, 2015b).

This increasing indifference towards climate change was underscored by the rise of economic and security concerns as well as scepticism and a sense of pessimism about the future of the European Union itself. Between 2011 and 2013, most Europeans felt that the worst impacts of the financial crisis were yet to come and public confidence in the EU's (rather than member state governments') ability to address these concerns was declining, as measured by the Commission's own polling (European Commission 2013a).

5.4 EUROPEAN COUNCIL, OCTOBER 2014

Although the European Commission had responded to some Member State pressure in creating its compromise proposals, there were several unresolved issues for the Heads of State and Government to discuss at the European Council in October 2014. A minority of members were

---

[1] In 2008, climate change was considered the most serious problem (European Commission 2008).

in favour of a GHG emission reduction target of 'at least' 40%, including Denmark and Germany as well as the UK. These members tend to have domestic targets for emissions reduction in place that was at least as ambitious as those proposed by the European Commission.

A larger group of member states had indicated that they could support a 40% GHG target, with some conditions. Support from Poland was conditional on exclusion from some of its energy sectors from the cost of ETS participation. Electrically isolated Portugal and, to a lesser extent Spain, and some of the Baltic nations, wanted to see the EU to mandate increased electrical interconnection between states to reduce electrical reliance on Russia (in the case of the Baltic states) and, to open up markets for excess renewable energy produced the Iberian countries. Of the 'at least' 40% GHG members, nearly all strongly supported ambitious targets for energy efficiency and renewable energy, largely in line with their domestic policy situation. The exception was the UK, which strongly preferred 'technology neutrality' in energy policies. The members that had indicated that they could support or accept an (only) 40% target had a far wider range of positions on the issues of renewable energy and energy efficiency but tended to favour less ambitious targets than, for example, Denmark and Germany.

Figure 5.1 gives an overview of the positions of all member states going into the European Council in October 2014. It can be seen that the UK was something of an anomaly, being strongly in favour of climate ambition while resisting multiple targets. In general, member states that favoured strong climate action also favoured multiple targets.

At the summit, Poland and the VISEGRAD nations, often portrayed as tough or even intransigent negotiators on the European Council, especially on climate issues (Interview 1, 2015; Janowska 2011), were able to use the threat of veto of the package to obtain concessions. These countries were granted an extension to the existing exclusion of parts of their power systems from the European Emissions Trading System (ETS) cap-and-trade scheme until 2030 and access to an energy system modernisation reserve based on ETS revenues valued at 300 million allowances between 2021 and 2030 (European Council 2014). In return, Poland was able to sign up to 'at least' 40% GHG reduction target. This was an outcome the Prime Minister declared as a 'win' for Poland (Euractiv 2014b). Portugal was able to bring interconnection onto the agenda but did not secure a binding target with an indicative 15% interconnection target agreed (Reuters 2014).

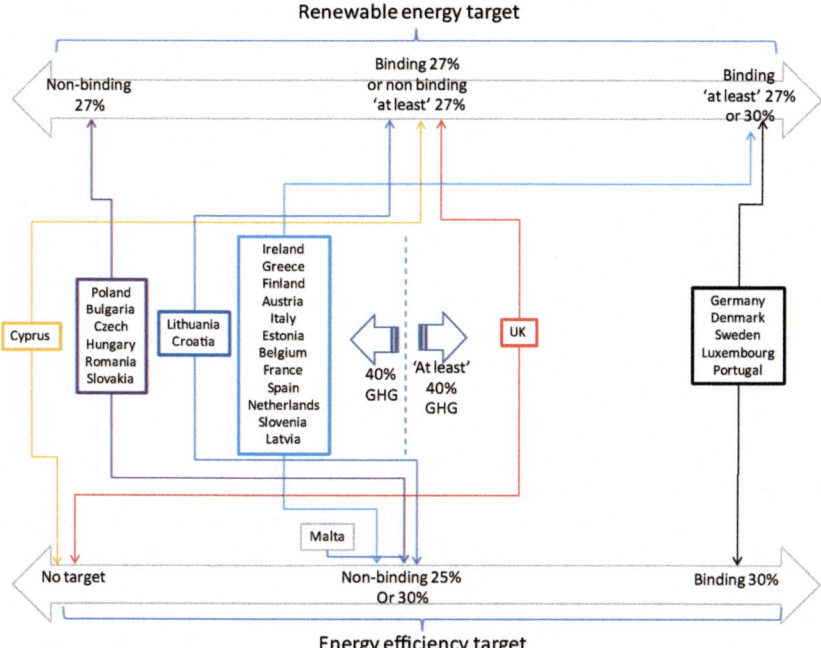

Fig. 5.1 Member state positions ahead of the October 2014 European Council (*Data source* Euractiv 2014a)

In summary, the outcome of the October 2014 European Council was an agreement relating to the four key targets the make up the 2030 climate and energy framework. To recap, these were:

- An EU target of reducing greenhouse gas emissions by at least 40% compared to 1990 by 2030. This target was to be shared between the non-ETS sector and the ETS sector sub-targets of which were 30 and 43% below 2005 levels respectively;
- A target of at least 27% of all energy to come from renewable sources, binding at the EU level with a new governance system announced to monitor and ensure progress towards the target;
- A reduction in energy demand compared to a business as a usual benchmark of 27%, with a review of progress at 2020 with an option to increase the 2030 target to 30%;

- A non-binding objective for all member states to achieve cross-border interconnection of 15% of their installed electricity production capacity by 2030 (although the guiding target remains 10% by 2020).[2]

5.5   Conclusion

MSA posits that developments in the political context, known as the politics stream, affect the policymaking agenda (Kingdon 2010, p. 145). This chapter has analysed those political events that were especially salient to the policymaking agenda for the EU 2030 targets, presenting a discussion of the broad currents that were evident in the European political sphere, such as it was, and the constellation of positions adopted by member states.

During the period 2011–2014, tackling climate change was waning in importance for many European voters with economic and security worries becoming more important. The cost of renewable energy support had become a difficult political issue in several EU member states. At the same time, the ability of the EU to address these problems was being questioned, contributing to a surge in support for broadly anti-establishment or anti-EU political parties at the 2014 Parliamentary elections.

The heterogeneity of positions within the European Council meant that even among those states favouring ambitious action on climate change, there was little agreement on the correct approach. Strong resistance to the policy from some countries exacerbated this lack of cohesion. Agreement on the final deal was only made possible by economic concessions made to Eastern and Central European Member States.

The response of the European Commission to these trends, while it drafted its proposals, was caution. Ambition for the 2030 package was determined on strictly economic grounds with much emphasis placed on the cost efficiency of policy choices as well as their implications for subsidiarity. The Commission appears to have been constrained by events in the political context, which placed downward pressure on the aspirations or extent of the targets as well as their scope, leading to proposals that tend towards emphasis of the EU-ETS rather that energy targets which Commission officials were felt to be more prescriptive.

---

[2]Although the Commission later warned that the 15% target may not be economically viable in all member states (European Commission 2015a).

The preceding three chapters have described the three streams of the MSA: namely the problem, policy and politics streams. MSA suggests that the convergence or 'coupling' of the three streams occurs only during time-limited 'policy windows', often under the influence of skilled actors known as policy entrepreneurs. The next chapter explains how and when the streams came together as well as describing the identity and role of policy entrepreneurs in that coupling.

REFERENCES

Bressanelli, E., Koop, C., & Reh, C. (2016). The Impact of Informalisation: Early Agreements and Voting Cohesion in the European Parliament. *European Union Politics, 17*(1), 91–113.

Crum, B. (2013). Saving the Euro at the Cost of Democracy? *Journal of Common Market Studies, 51*(4), 614–630.

Euractiv. (2014a). *Member States' Positions on 2030 Climate and Energy Targets Revealed.* Euractiv.com. Available at: http://www.euractiv.com/sections/energy/member-states-positions-2030-climate-and-energy-targets-revealed-309279. Accessed 13 Apr 2016.

Euractiv. (2014b). *Poland Says It 'Won' at the EU Summit.* Euractiv.com. Available at: http://www.euractiv.com/sections/energy/poland-says-it-won-eu-summit-309494. Accessed 24 Nov 2014.

European Commission. (2008). *Eurobarometer Special Report: Climate Change 2008.* Available at: http://ec.europa.eu/commfrontoffice/publicopinion/archives/ebs/ebs_300_full_en.pdf. Accessed 3 May 2016.

European Commission. (2009). *Eurobarometer Special Report: Climate Change 2009.* Available at: http://ec.europa.eu/commfrontoffice/publicopinion/archives/ebs/ebs_313_en.pdf. Accessed 3 May 2016.

European Commission. (2011). *Eurobarometer Special Report: Climate Change 2011.* Available at: http://ec.europa.eu/commfrontoffice/publicopinion/archives/ebs/ebs_372_en.pdf. Accessed 3 May 2016.

European Commission. (2013a). *Europeans, the European Union and the Crisis: Standard Eurobarometer 79 Spring 2013.* Available at: http://ec.europa.eu/commfrontoffice/publicopinion/archives/eb/eb79/eb79_first_en.pdf. Accessed 4 May 2016.

European Commission (2013b). *Report on the Implementation, Results and Overall Assessment of the 2013 European Year of Citizens.* Available at: http://europa.eu/citizens-2013/sites/default/files/content/document/COM_2014_687_F1_REPORT_FROM_COMMISSION_EN_V4_P1_789180.PDF. Accessed 6 May 2016.

European Commission. (2014). *Eurobarometer Special Report: Climate Change 2014.* Available at: http://ec.europa.eu/commfrontoffice/publicopinion/archives/ebs/ebs_409_en.pdf. Accessed 3 May 2016.

European Commission. (2015a). *Connecting Power Markets to Deliver Security of Supply, Market Integration and the Large-Scale Uptake of Renewables.* Available at: http://europa.eu/rapid/press-release_MEMO-15-4486_en.htm. Accessed 19 May 2016.

European Commission. (2015b). *Eurobarometer Special Report: Climate Change 2015.* Available at: http://data.europa.eu/euodp/en/data/dataset/S2060_83_4_435_ENG. Accessed 3 May 2016.

European Council. (2014). *European Council (23 and 24 October 2014) Conclusions on 2030 Climate and Energy Policy Framework.* Available at: http://www.consilium.europa.eu/uedocs/cms_data/docs/pressdata/en/ec/145397.pdf. Accessed 24 Oct 2014.

Herweg, N., & Zahariadis, N. (2018). The Multiple Streams Approach. In N. Zahariadis & L. Buonanno (Eds.), *The Routledge Handbook of European Public Policy* (pp. 32–42). Oxford: Routledge.

Hurrelmann, A. (2007). European Democracy, the 'Permissive Consensus' and the Collapse of the EU Constitution. *European Law Journal, 13*(3), 343–359.

Janowska, K. (2011). Poland's Climate Change Policy Struggle. In R. Wurzel & J. Connelly (Eds.), *The European Union as a Leader in International Climate Change Politics.* Abingdon: Routledge.

Kingdon, J. W. (2010). *Agendas, Alternatives, and Public Policies* (2nd ed.). Harlow: Pearson.

Lane, P. R. (2012). The European Sovereign Debt Crisis. *Journal of Economic Perspectives, 26*(3), 49–68.

Lapavitsas, C., et al. (2010). Eurozone Crisis: Beggar Thyself and Thy Neighbour. *Journal of Balkan and Near Eastern Studies, 12*(4), 321–373.

Lindberg, L. N., & Scheingold, S. A. (1970). *Europe's Would-Be Polity: Patterns of Change in the European Community.* Englewood Cliffs: Prentice-Hall.

Marks, G., & Hooghe, L. (2009). A Postfunctionalist Theory of European Integration: From Permissive Consensus to Constraining Dissensus. *British Journal of Political Science, 39*(1), 1–23.

Princen, S. (2007). Agenda-Setting in the European Union: A Theoretical Exploration and Agenda for Research. *Journal of European Public Policy, 14*(1), 21–38.

Reuters. (2014). *Portugal Could Block EU Climate Deal Over Connection Target.* Available at: http://www.reuters.com/article/eu-summit-climatechange-portugal-idUSL6N0SH1KI20141022. Accessed 18 May 2016.

Spiegel, P., & Carnegy, H. (2014). *Anti EU Parties Celebrate Election Success.* FT.com. Available at: https://next.ft.com/content/783e39b4-e4af-11e3-9b2b-00144feabdc0. Accessed 5 Nov 2015.

The Economist. (2014). *The Eurosceptic Union*. Economist.com. Available at: http://www.economist.com/news/europe/21603034-impact-rise-anti-establishment-parties-europe-and-abroad-eurosceptic-union. Accessed 5 Nov 2016.

Torreblanca, J. I., & Leonard, M. (2013). *The Continent-Wide Rise of Euroscepticism*. London: ECFR.

Van de Steeg, M. (2006). Does a Public Sphere Exist in the EU? An Analysis of the Content of the Debate on the Haider Case. *European Journal of Political Research, 45*, 609–634.

VoteWatch Europe. (2015). *Who Holds the Power in the New European Parliament? And Why?* Available at: http://60811b39eee4e42e277a-72b421883bb5b-133f34e068afdd7cb11.r29.cf3.rackcdn.com/2015/02/VoteWatch_template_web.pdf. Accessed 10 Jan 2017.

Zahariadis, N. (2007). Multiple Streams Framework: Structure, Prospects, Limitations. In P. A. Sabatier (Ed.), *Theories of the Policy Process* (pp. 65–92). Boulder, CO: Westview Press.

Zahariadis, N. (2008). Ambiguity and Choice in European Public Policy. *Journal of European Public Policy, 15*(4), 514–530.

CHAPTER 6

Connecting the Streams

Abstract For policy change to occur, a policy window must open to allow a policy entrepreneur to connect their preferred solution to a salient problem. This chapter shows that, while a policy window did open in 2013 and 2014, it was narrower and harder for policy actors to navigate than the unambiguous opportunity for change that was present in 2007. Within this complex and unpredictable environment, none of the observed attempts at entrepreneurship were unqualified successes, despite some notable achievements.

Keywords Policy entrepreneurship · Policy windows · Advocacy coalitions

6.1 Introduction

The three previous chapters described the three streams of the multiple streams approach (MSA). This chapter now proceeds by describing two other important elements. The first is that an appropriate 'policy window' needs to be open. In other words, that the time is right for *'advocates of proposals to push their pet proposal or conception of a policy problem'* (Kingdon 2010, p. 165; Zahariadis 2007). The second aspect relates to the action of policy entrepreneurs who seek to combine the three streams.

© The Author(s) 2018   99
O. Fitch-Roy and J. Fairbrass, *Negotiating the EU's 2030 Climate and Energy Framework*, Progressive Energy Policy, https://doi.org/10.1007/978-3-319-90948-6_6

MSA suggests that only during special moments or 'policy windows' can advocates 'couple the streams' or push their ideas onto the agenda with any success (Kingdon 2010). This chapter explains the coupling process, the timing of the policy window opening and the role played by policy entrepreneurship in the process.

This chapter shows that:

1. Not all actors experienced the same policy window, with opportunities to act being presented to different actors at different times. The nature of the policy window also changed over time in response to events in the politics and problem streams;
2. Several groups or individuals can be said to have exhibited at least some of the qualities of a policy entrepreneur; and
3. Ideas and networks 'spiltover' in the 2030 debate from the contemporaneous debate about ETS reform with significant implications for the 2030 policy.

Following short summaries of the concepts of 'policy windows' and 'policy entrepreneurs' in Sect. 6.2, this chapter has three main sections. Section 6.3 describes the policy window within which the 2030 climate and energy framework was negotiated. Section 6.4 presents evidence of entrepreneurship and identifies actors who may be usefully described as having acted in an 'entrepreneurial' way in the lead up to the setting of the 2030 climate and energy targets. Section 6.5 concludes.

6.2   MSA, Policy Windows and Policy Entrepreneurs

This section recaps the main point relating to MSA, policy windows (Sect. 6.2.1) and entrepreneurs (Sect. 6.2.2) discussed earlier in Chapter 2.

6.2.1   Policy Windows in Multiple Streams

In MSA, a policy window is a moment in time when there is '*opportunity for action*' (Kingdon 2010, p. 166) for advocates of a particular policy outcome. An open policy window is a prerequisite for policy change and it provides any group with an interest in policymaking with an opportunity to influence that change. In his original formulation of MSA, Kingdon suggested that policy windows occur infrequently and stay open for only a short period of time although policy windows have since been shown, in some settings, to remain open for years at a time (Carter and Jacobs 2014). However long the duration, actors must be ready to take advantage of the window while it is open or face having to wait for the opportunity to present itself again. Windows may open in either the politics stream or the problem stream due to focusing events or disasters, personnel changes due to elections, or due to cross-pollination of ideas and activity from related policy areas known as spillover (Ackrill and Kay 2011; Kingdon 2010).

Windows can open predictably, such as when time-limited legislation expires or budgets are approved or they may be '*as unpredictable as earthquakes*' (Zahariadis 2007). The metaphor of surfers waiting for waves has been used to describe the concept. Policy advocates must be prepared, must have their ideas, messages and networks ready to catch the 'big wave' when it comes. Attempts are sometimes made to catch the wrong wave but eventually the wave breaks that will bring them to shore (Kingdon 2010, p. 165; Boscarino 2009).

6.2.2   Policy Entrepreneurs in Multiple Streams

In MSA, policy entrepreneurs are the skilled, resourceful actors who attempt to bring the three streams together: to couple their preferred policy idea to salient policy problems in a way that is acceptable in the prevailing political context. Entrepreneurs are an important factor in explaining policy change, but at least one study has shown that they are not essential in all contexts (Herweg 2016). The second half of this chapter, Sect. 6.4, seeks to confirm whether policy entrepreneurship occurred during the negotiation of the 2030 targets and, if so, the identity of any entrepreneurs who acted and their influence on the agenda.

In a world where problems are not solved in a linear, rational manner, far from simply being advocates of particular policy solutions, policy entrepreneurs must be 'power brokers' and manipulators of ambiguity in order to '*craft contestable meaning which they, in turn, disseminate to policymakers in order to activate attention and mobilize support or opposition*' (Ackrill et al. 2013, p. 873). Entrepreneurs can be from the corporate, non-governmental or policymaking spheres. Their location is '*almost irrelevant*' but they are always '*central figures in the drama*' (Kingdon 2010, p. 180).

Invariably, there are many actors who contribute and review a given policy decision. These actors come from inside and outside the formal policymaking institutions and many have a material interest in causing, directing or preventing policy change.

Successful policy entrepreneurs, as seen by MSA (Kingdon 2010) tend to have three identifying qualities. First, they must be credible or, at least, must have some claim that they should be heard. This may be the product of particular expertise or specialist knowledge. It could be the ability to speak on behalf of others, or it could be some power over the decision-making process, as held by officials and policymakers. Second, entrepreneurs have uncommonly good negotiation and networking skills and they are renowned for them. Third, entrepreneurs must be tenacious. They 'put in the time' and are there at the meetings, on speaking panels, writing positions papers and, as one interviewee put it, *'generally keeping up the noise levels'* and expending *'blood, sweat and tears'* (Interview 8, 2015; Kingdon 2010).

More generally, policy entrepreneurs tend to display the following characteristics:

- **Social acuity** or perceptiveness and understanding others when engaging in policy conversations. This is generally demonstrated as making good use of policy networks and by *'understanding the ideas, motives, and concerns of others and responding effectively'*;
- **Defining problems** in ways that influence who pays attention to them;
- **Building teams** or the ability to work effectively with others by, for example, building coalitions of people with complementary skills; and
- **Leading by example** by, for example, turning an idea into action to introduce or demonstrate the concept and overcome the reservations of generally risk-averse policymakers and *'signal a genuine commitment to improved outcomes'* (Mintrom and Norman 2009, p. 653).

The overriding objective of a policy entrepreneur is to create a package of problem, policy and politics which can then be 'sold' to policymakers (Ackrill et al. 2013). Entrepreneurs, however, in the same way as any other political actor, have limited financial, reputational and cognitive resources available to them. They may have more than one set of demands on their abilities to form networks and coalitions and to connect problems and solutions.

While preparedness is a powerful asset for any policy entrepreneur who must *'keep the gun loaded'* in case an opportunity arises, the importance of serendipity in an essentially chaotic system should not be overlooked. An advocate may experience multiple failures of a strategy, attempting to push the wrong solution for the wrong problem under the wrong political conditions. But, on another day, in another policy window, the same idea may have a radical impact on policy.

Kingdon (2010, p. 182) points out that the concept of policy entrepreneurs and policy windows helps to make sense of the age-old structure-versus-agency debate (Ackrill et al. 2013). A window opens for structural reasons beyond the agency of any individual while making advantage of that opportunity requires an individual to act with agency.

Entrepreneurs are able to read the political and policy landscapes, spot opportunities and forge important connections between actors. The role of policy entrepreneur described by MSA is, fundamentally, a creative one. Entrepreneurs have excellent political instincts and tend to make the most out of their resources through the ability to harness political forces beyond their control (Kingdon 2010, p. 181). Entrepreneurs are also adaptable and able to compromise when needed. While many advocates may adopt radical or controversial policy positions during normal times, when—and only when—a policy window opens, entrepreneurs bargain, make connections and give concessions in order to maximise their impact (Kingdon 2010).

6.3   The 2030 Policy Window(s)

In addition to the momentum from the 2005 'Hampton Court Speech' in which UK Prime Minister, Tony Blair effectively rebooted EU climate and energy policy, in 2007, there was a strong push from the French EU presidency to make progress. The potential for policy change represented a clear and exploitable policy window (Boasson and Wettestad 2013). This time around, however, the various presidencies (Ireland, Lithuania, Greece) leading the Council did not make an ambitious climate and energy package a priority in the same way. Compared to the policy debate that led to the 20/20/20 package in 2007, the political context of the 2030 package was much more complicated and less conducive to ambitious change. The financial crisis, a more challenging energy security situation, a new Directorate-General for Climate Action (DG Clima) with a remit to strengthen the EU-ETS and growing popular mistrust of European Union decisions all added to the complexity.

The '2030 debate' tested the skills of the most talented climate and energy lobbyists, the political terrain being especially '*hard to read*' (Interview 13, 2015). If a policy window opened, it was certainly narrower and more difficult to navigate than the one that produced the 20/20/20 package in 2007. Nevertheless, we argue that a policy window did open and that political actors did exploit it, resulting in an observable effect on the policy agenda in 2014.

The following sections look at the timing of the window opening and closing in Sect. 6.3.1, and some of the important characteristics of the policy window in Sect. 6.3.2.

6.3.1   A Policy Window Closes

The metronome of the quarterly European Council meetings of Heads of State and Governments sets the tempo of European political life. Policy windows are opened and often closed by the timing of these summits (de Schoutheete 2012). In the case of the 2030 climate and energy framework, identifying the closure of the policy window is relatively straightforward since there was a clear policymaking deadline. The October 2014 summit was widely anticipated to produce a decision on the level and structure of the EU's climate and energy goals after 2020 and therefore, mark the closure of the policy window. The timing of this particular summit was important, as it was seen to be the last opportunity for the EU to agree on a climate target in time to conform to the timeline set out by the United Nations in preparation for the COP 21 climate talks in Paris at the end of the following year (European Council 2014). The Commission's policymaking timetable worked backwards from this point.

The timing of the opening of the policy window, however, is slightly less clear with three policymaking cycles interacting: the 20/20/20 package expiry, reform of the EU-ETS and the UN climate change negotiations timetable. As early as 2008, when the 20/20/20 package was agreed, it was clear that the package had a built-in expiry date and that, at some point in the future, some form of policy action would be needed which addressed the period after 2020. Ahead of COP 15 in Copenhagen in 2009, the European Council had committed the Union to 80–95% reductions in GHG emissions, compared to 1990, by 2050.

In late 2011, one of the follow-up UN climate conferences to the Copenhagen conference, this time in Durban, South Africa, set a date for a '*universal legal agreement on climate change no later than 2015*'. The conference also decided the timetable for future COP meetings with a conference planned for December 2015, although the host city, Paris, was not announced until later (UNFCCC 2012).

Therefore, by the end of 2011, actors who were familiar with both the timetable of the UN climate talks and the deadlines implicit in the EU's own climate and energy policy programme may have been able to predict that decisions about a new EU climate and energy policy were likely to be forthcoming. They may also have noted that the post-2020 policy, whatever it was likely to be, may have to be accelerated to enable the EU to contribute to the important international climate talks in 2015.

A policy window opened slightly at around the time of COP 15 in late 2009 but at this early stage, very few actors were able to take advantage of it. It was difficult for most civil society actors to gauge the range of likely payoffs of pushing their ideas or attempting to manipulate political events. Two actors, however, who were able to exploit the inevitability of some kind of EU-level policy to replace the time-limited 20/20/20 package, were the European Climate Foundation (ECF) and Eurelectric. They were able to work with the European Commission to begin thinking about the technical and economic implications of a climate and energy policy that would reach the European Council's stated aim of an 80–95% reduction in greenhouse gas emissions by 2050. ECF, as an environmental foundation, was able to participate for two reasons. First, its commitment to '*the cause, not the brand*' meant it was actively searching for opportunities to engage 'behind the scenes' and fill knowledge gaps rather than the more partisan advocacy seen later on in the process. Second, within the environmental community, it was most able to put significant financial resources at risk with no guarantee that it would translate into policy. Eurelectric, on the other hand, was more timid in its activity with a single, less expansive or influential report (Eurelectric 2009).

As it became clearer, partly due to the analysis undertaken by ECF, that longer-term issues of decarbonisation were both complex and, from a policy perspective, not well understood at the EU level, more actors saw an opportunity to contribute modelling analysis and DG Energy producing the 'Energy Roadmap 2015' at the end of 2011.

The highly technical outputs of the modelling work, especially by ECF, opened the policy window wider, emboldening several advocacy groups to produce a flurry of modelling reports about how to decarbonise by

2050. By 2013, the Commission had connected the 2050 target with the expiry of the 20/20/20 package, leading to a year of 'targetology' with various formulations of targets proposed and countered within the policy. Eventually, the window began to close as the 2015 deadline of the Paris COP approached. The Commission made its proposal in early 2014, which shifted attention to the European Council, which was to take a decision in October that year. Energy Efficiency and the nature of a target was somewhat muted during the targetology phase and absent from the January 2014 proposals due to an ongoing policy review within the European Commission. By the time proposals for an energy efficiency target were made in July 2014, the policy window expanded slightly due to the escalating energy dispute between Russia and Ukraine

Other than a hard-core of especially engaged experts, most civil society actors were effectively powerless to act until the Commission, more or less on its own initiative and in keeping with the timeline, rather than in response to a direct request by the European Council, formally opened up the discussion with a Green Paper in Spring 2013. The timing of the paper itself was ultimately an artefact of the international climate negotiations timeline and the impending expiry of the 20/20/20 framework (European Commission 2013).

Indeed, Commission officials responsible for the package only acknowledge that a discussion about the 2030 targets began in 2013 and some lobbyists first became aware of the opportunity to act during the drafting of the green paper (Interview 25, 2015).

6.3.2   Nature of the Policy Window

That the EU policy for after 2020 was seen as an opportunity by civil society actors and the Commission to shift the course of climate and energy policy was the result of recent developments in the problem and politics streams. In the problem stream, the economic crisis and worries about the cost of energy to firms and households had made cost effectiveness a primary concern for all policymakers, regardless of the policy area. At the same time, stemming from declining trust in the EU as well as the poor outcome from Copenhagen climate conference, there was an opportunity to make arguments against ambitious EU action, especially action that risked the EU moving too far ahead of the international community on climate. Towards the end of the window's opening, events in

Ukraine briefly widened it. The topic of energy security, which had been an important but largely background issue, suddenly leapt into the foreground in June 2014 when Russia cut off the gas to Ukraine, threatening European energy security.

Even after the European Commission published its Green Paper, for some topics, making progress on the agenda was hindered by the timetable. The energy efficiency portion of the discussion, in particular, remained somewhat sidelined due to an ongoing commitment contained in the 2012 Energy Efficiency Directive (EED) for the Commission to review energy efficiency progress annually, postponing serious discussion about proposed targets for 2030 until summer 2014 (European Commission 2012).

Overlapping Windows

As described earlier, legislation aimed at reforming the EU-ETS was in progress at the same time that the 2030 targets were a live policy issue. A clear EU-ETS policy window was open, largely due to the extremely low price of allowances in the trading system. From the perspective of the climate and energy policy community, the negotiation of the EU 2030 targets and EU-ETS reform were perceived as either part of a continuous process or as very closely linked. One interviewee pointed out that:

> Remember, all of this [*2030 discussion*] is overlapped with the ETS back-loading battle on the beginning of the ETS market stability reserve battle and the discussion about whether the carbon price would ever get beyond its ridiculously low level. (Interview 8, 2015)

The temporal overlap between the legislative phase of the ETS reform debate and the agenda-setting phase of the EU 2030 debate was important for the 2030 targets discussion in several ways.

First, most civil society actors with an interest in climate and energy policy were spending at least some time and resources between 2010 and 2013 lobbying for or against ETS reform, limiting their ability to engage wholeheartedly in the 2030 debate, at least in the early stages. Second, significant coalitions such as 'Friends of ETS' were formed to exploit the ETS reform window, which persisted into the 2030 policy window. Finally, the coalition of pro-ETS reform actors framed 'fixing the ETS' as the primary task faced by EU climate and energy policy.

6.4   Policy Entrepreneurs

The previous section sets out the opening, closing and nature of the policy window. This section explores whether or not any civil society actors engaged in the 2030 debate display the qualities of policy entrepreneurship. First, we discuss some of the difficulties of using interview data obtained from participants who self-identify as entrepreneurs in Sect. 6.4.1. The section then takes each of the characteristics of policy entrepreneurs proposed by Mintrom and Norman (2009) and presents evidence relevant to each of them. Subsection 6.4.2 looks at social acuity, 6.4.3 looks at how actors define problems, 6.4.4 at team building and 6.4.5 at leading by example.

6.4.1   Problems with Self-Identified Entrepreneurs

Identifying policy entrepreneurs can be problematic. Many civil society and business actors are under pressure to demonstrate to supporters, funders or employers that they are able to exert influence on policy. In the interview, we observe a bias towards self-reported successes, so-called 'expansiveness bias' (Beyers et al. 2014, p. 179). Lobbyists and campaigners are strongly inclined to make a case that their strategy was not only successful, but that it, and their political skill, was superior to that of others.

To mitigate this self-assessment problem, we take two measures. First, we add a more careful specification of policy entrepreneurship than proposed by MSA that helps make our assessment of entrepreneurial activity more objective. Our assessment of policy entrepreneurship is based on the qualities suggested by Mintrom and Norman (2009): social acuity, problem definition, team building and leading by example, is used to identify entrepreneurship. We expand on Mintrom and Norman's specification below. Second, we rely wherever possible on third-party accounts of actors' performance, especially instances of self-reported success.

6.4.2   Social Acuity

There are two dimensions to social acuity. The first is *getting the timing right*, or the ability to read the policymaking context in order to appreciate when a policy window is open or is about to open. The second is *making friends and influencing people* or the ability to make critical interpersonal connections since 'being well connected' and 'getting along well with others', even those with whom you disagree (or perhaps especially) are important if one wants to achieve anything in the policy community (Mintrom and Norman 2009). The following subsections look at each of these dimensions in turn.

Getting the Timing Right
Lobbyists tend to have a keen sense of the agenda-setting process and the changing nature of opportunities to engage through time. The environmental group, E3G, for example, uses a conceptual tool it describes as the 'decision-funnel' to describe how a debate moves from generalities, through a policy debate and on to a specific choice between a small selection of options (E3G 2014). Brussels lobbyists generally try to respond to an opportunity as soon as an accessible policy window opens. While there may be some *'space to influence'* later on, *'for us, the time to engage is early in the process'* (Interview 25, 2015). It is also important for all actors to recognise that attention to what they are saying is a limited resource and that they need to try and time their interventions well. As one lobbyist put it:

> I guess you have to choose your moment when you're going to fight because if you're just noisy all the time, you're just a noise and people ignore you. (Interview 27, 2016)

As discussed in the previous section, the policy window opened slightly in late 2009 around the time of the Copenhagen COP 15 climate conference. At this time, however, very few actors from civil society were able to become involved in the process. Those that were, primarily Eurelectric and ECF, seem to have spotted what others had not: that some form of follow-up package to the 20/20/20 deal (which had only been legislated that year) was inevitable if the EU was to take its own climate ambition, as decided by the European Council ahead of Copenhagen, seriously.

That ECF and Eurelectric were able to engage at this early stage was as much due to their characteristics as their reading of the policy landscape. ECF was well regarded and connected, and had access to the funds, to produce a detailed (and presumably expensive) piece of analysis with no firm sign of a payoff yet on the horizon. This kind of high-risk speculative activity might be hard to defend for most trade associations or ad hoc groups to justify funding, even if they were able to anticipate the policy direction.

From 2011 onwards, advocates of energy efficiency policy, led by EuroACE, had been working to associate the 'solution' of greater energy efficiency with energy security (Warren 2011). These efforts were escalated in 2014 as the Ukraine-Russia energy dispute intensified with the Secretary-General of EuroACE able to present the ideas to an informal meeting of Europe's energy ministers at the moment the crisis reached its height (Greek Presidency of the Council of the European Union 2014).

Making Friends and Influencing People

Entrepreneurs tend to '*understand the ideas, motives, and concerns of others*' (Mintrom and Norman 2009). BUSINESSEUROPE and the energy-intensive industries were one group that the 2030 debate revealed to be especially reluctant or unable to deploy these kinds of soft-skills. BUSINESSEUROPE's style of political engagement was seen by others, both inside and outside the formal policy process, to be 'aggressive' and often 'confrontational'. Some point out that BUSINESSEUROPE's strategy on climate and energy lobbying is to consistently point out problems such as the costs of certain policy choices for business rather than signal a commitment to problem-solving in partnership with policymakers and are often seen as arrogant or at least overly direct (Interview 1, 2015; Interview 29, 2016; Interview 30, 2016). Some observers believe that weakness of these industries to engage on a social level with peers and policymakers might be explained by their assessment that a large proportion of employees in the energy-intensive industries are middle-aged men, less comfortable in settings that are more diverse:

> I have to say I think a lot of that was also to do with age group and gender. So inside the BUSINESSEUROPE working group on climate change, there were, not infrequently, incidents where a man in his 50s representing one of the heavy industry type interests would say, 'Oh, another of you bloody women,' because actually, as it happened there were about five younger women who attended that working group and we all worked on the [opposing] side of the issue. (Interview 8, 2015)

By contrast, the Friends of ETS was well regarded by other lobbyists. People from the electricity sector, utility companies and even BUSINESSEUROPE, who fiercely opposed the policy positions taken by the coalition, described key figures in the coalition in favourable terms. For example, political adversaries were able to describe them as *'great'* or as having *'... a very, very good overview, and very great insights'* (Interview 8, 2015; Interview 15, 2015; Interview 18, 2015; Interview 26, 2016). Some participants put the favourable assessment down to trust, the ability to *'gee up the troops' and 'keeping people together'*:

> You know, [*they*] have no traction whatsoever, other than the fact that [*they are*] good at spotting opportunities, good at organising things, and good at getting people to work together. (Interview 30, 2016)

The Magritte Group's high-profile members were able to make connections with senior national and European policymakers. The group secured meetings with the head of state or government of France, Germany and the Czech Republic as well as Commissioners including Günter Oettinger and Parliamentary leaders (CEZ Group 2014; ENGIE 2014). The conception of the group was to engage at a *"senior level"* and senior policymakers apparently welcomed the engagement with their élite peers from the world of business (Interview 1, 2015; Interview 3, 2015).

6.4.3 Defining Problems

Different actors framed the problems facing EU climate and energy policy in various ways. The energy-intensive industries continued their strategy of making 'carbon leakage' a primary problem not only of the EU-ETS, but also of climate and energy policy more generally, as they had for many years. While this strategy had proven successful in the past, gaining a 'carbon leakage list' and other concessions, the energy intensives and BUSINESSEUROPE had recently strongly but unsuccessfully resisted ETS reform, leading to some reputational damage (Wettestad and Jevnaker 2016).

> During the ETS back-loading row, they were very emphatically and very publicly hostile and that burned them. Not only did they lose the battle but they also got a lot of reputational burn. (Interview 8, 2015)

Despite the ETS reform setback, the economic crisis in Europe somewhat enhanced the perception of their arguments for taking Europe's industrial competitiveness seriously (Interview 4, 2015). Many of the arguments put forward in favour of protecting heavy industry had, over the years, been well absorbed and understood by policymakers in the Commission who see little new constructive thinking coming from that quarter (Interview 17, 2015).

The energy-intensive industries also worked to show that unilateral climate action, such as setting a GHG emissions reduction target ahead of the talks in Paris, would be damaging to Europe's international competitiveness. An approach that one lobbyist described as;

> … a sort of St Augustine position, you know, "We can't say no to climate policy but we don't really want it. Lord, make me chaste but not yet." It was always the tomorrow maybe in the future, non-definite clause thrown into anything positive they said. (Interview 8, 2015)

In its efforts to secure ETS reform, the Friends of ETS worked hard to frame Europe's climate policy problem as a failure of the emissions trading system. The group made sure that:

> Every single bit of coverage leading up to any communication from the Commission [on energy and climate] we said, 'We have a crisis in the ETS.' (Interview 30, 2016)

For members of the Friends of ETS coalition, this posed a problem:

> "We couldn't talk about multiple targets, which meant we couldn't get the wind people on board"… "the wind people wanted to keep on putting in language saying 'multiple targets'. We kept on saying 'it's only the ETS that counts'"…"[so]this was the sacrifice we had to make. It was either to get wind [*energy advocates*] on, or we can go and get more people who are ultra-the-heart-of-darkness, but quite frankly they've got more power, and [*we would*] rather talk to the guys who've got more [*power*]." (Interview 30, 2016)

The energy efficiency trade association, EuroACE along with NGOs such as Friends of the Earth worked hard for several years to emphasise the problem of European energy import dependence with their 'energy dependence day' campaign, which was accelerated during the Ukraine-Russia gas dispute of 2014.

The Magritte Group worked hard to frame the problem as a security one of electricity generation adequacy, caused by rapid expansion of renewables, largely because of the 20/20/20 framework legislated in 2009. At the same time, with some success, they helped to push the problem of increasing consumer energy bills into the conversation, also framed as a problem caused by adding renewables to the mix, a line that was supported strongly by organisations such as Shell (Interview 25, 2015). The Magritte group was also very aware that European politicians, especially those in the Commission, were nervous about declining trust in the EU and its institutions. In response, the group was very clear in defining all the problems as 'European problems' that would require 'European solutions', a framing that was well appreciated by Commissioners, particularly when the Magritte group delivered it in person to Heads of State and Governments of member states, in whose hands the future of European integration ultimately lay (Interview 1, 2015).

The ECF had perhaps the most sophisticated or, at least, best funded, approach to problem definition. Rather than relying on existing rhetoric, ECF sought to introduce entirely new knowledge to the debate with their Roadmap 2050 project. The project, which looked at the problems associated with decarbonisation, was an important milestone in setting the terms of reference for what would become the 2030 debate. At a very early stage, it successfully established that renewable energy, energy efficiency and improving energy infrastructure were 'no regrets' options for decarbonising the electricity sector.

6.4.4   Team Building

As we discussed in Chapter 4, several alternative groupings were active in the debate about the 2030 targets. Coalition building and teamwork are a primary activity of civil society actors in Brussels and there are three main reasons for doing it (Mintrom and Norman 2009). These are:

1. Demonstrating that not only is there wide support for an idea among a large number of companies or organisations, but that the support is broad-based and actors with diverse interests are in favour of it. Coalitions including 'strange bedfellows', 'unholy alliances' or 'coalitions of the unlikely' of interests that would not

normally take the same position are seen as especially potent since they signal that the idea or position has wider support (Fairbrass 2013; Monciardini 2016; Beyers and De Bruycker 2017);

2. Intelligence gathering and sharing. A coalition of actors is a valuable means of sharing information and intelligence; and

3. Division of labour. A coalition is a good way of gaining access to the wide range of the resources needed to mount effective lobbying campaigns (Interview 1, 2015).

The following three subsections look at each of these rationales for team-building in turn.

Demonstrating Broad Support

The Magritte group of energy utilities consisted of ten or so very similar organisations but those organisations spanned a large geographical area with its members all prominent energy suppliers and employers across much of the European continent[1]. Altogether, the group was able to claim that it provided a half of Europe's electricity, serviced hundreds of millions of customers and employed two-thirds of a million people (Magritte Group 2013). Finding common ground between companies with very similar business models was not a great challenge for instigator Gérard Mestrallet of GDF Suez (now Engie), especially when that business model faced, by some reckoning, a common existential challenge, as described in Chapter 3. There was, however, some friction within the group over the topic of additional remuneration support for what the group described as under-used electricity generation capacity leading to the departure of Swedish Utility Vattenfall from the group in 2014 (Interview 1, 2015; Crouch 2014). The sheer political mass of the Magritte group enabled it to access decision-makers at the highest levels with meetings scheduled with powerful figures in all the EU's institutions including Commissioners and leaders of member state governments.

On the topic of the GHG emissions target, by early 2014 it seemed to some that those favouring low climate ambition (35%) and those arguing for more ambition (40% or more) had reached a stalemate and were

---

[1] The group included major gas and electricity suppliers from France, Germany, Spain, Czech Republic, Finland, the Netherlands and Italy. The member from Sweden, Vattenfall, left the group in early 2014 (Crouch 2014).

effectively *'cancelling each other out'* (Interview 8, 2015). In response, some sought to introduce actors to the community that were not usually directly involved in debating EU climate and energy policy.

> ...we teamed up with [*non-energy companies*], to make clear to policymakers that it's going broader, the coalition goes broader than just the energy sector. (Interview 18, 2015)

For example, the Institutional Investors Group on Climate Change (IIGCC), representing investors holding trillions of Euros in assets, rarely engages with the EU policymaking process beyond responding to consultations by the Commission. But, largely coordinated by a lobbyist working for electricity association, Eurelectric, the IIGCC members visited Brussels in January 2014 and met with several European Commission ers to make a case for a 40% target on behalf of the finance community. As a result of this experience, the IIGCC significantly increased its efforts on EU climate policy (Interview 8, 2015; Blanqué et al. 2015).

Within the renewable energy sector, the slow demise of the European Renewable Energy Council (EREC) through 2013 and 2014, following a real-estate dispute, made team building within the renewable energy sector much harder. As one frustrated renewables advocate put it:

> In terms of the Brussels debate, the absence of EREC has penalised us and is still penalising us, to an extent, because you cannot consult fifteen small associations every time you need to make a decision when you have Eurelectric, Eurogas and then a plethora of, "I might have 10 megawatts", "Oh I have 1000 megawatts but they're only in one country". No one cares. Get together, sort it out and then come with one composition. And that's what EREC could do. (Interview 13, 2015)

Other groups relied on building teams with representation from across different sectors and interest groups. For example, the Prince of Wales Corporate Leaders Group's inclusion of prominent members from a wide range of industries was an attempt to speak on behalf of as wide a business constituency as possible (Interview 29, 2016). The weight of the brands under the umbrella, as well as the endorsement of the Heir to the British Crown, opened doors for the coalition with a twice-annual meeting organised with Commission President Barroso (Interview 29, 2016).

Some of the sharpest lines of division within the policy community were among businesses and ENGOs and between renewable energy companies and incumbent energy firms. Relatively few teams or coalitions managed to bridge these divides. The Coalition for Energy Savings, for instance, includes members from NGOs and companies with a strong common interest in promoting energy efficiency. Trade associations such as the International Emissions Trading Association (IETA) include diverse actors, all of whom stand to benefit from expanded use of emission trading. The 'Friends of ETS' coalition, however, was able to include conventional energy companies, renewable energy associations, gas and electricity trade associations, oil companies and chemical firms in the same coalition with a less clear common interest in ETS reform. The coalition, coordinated by a small environmental NGO, clearly saw that having a range of voices speaking in favour of an idea, especially from the business world, was important for getting a point across:

> [*If we*] go in and say, 'Hey, Commission, we need to change the ETS target...' [*they will ask me to leave*]. But, if the likes of Shell and [*Italian utility*] Enel go in and say, 'We represent X million jobs and this much investment, and we think it's an important thing...' we know that the business orientated institutions will listen to them more than they'll listen to us. (Interview 30, 2016)

In order to create a coalition like 'Friends of ETS', its leaders needed to overcome a degree of scepticism on both sides of the divide but especially among the ENGOs:

> "Some of the NGOs were quite wary of having businesses move into the 'green space' because a lot of ethos for the historical origin of those NGOs is that this should be about social outcomes, not about business outcomes" ..."there were definitely people in the NGO world who felt [*Friends of ETS*] was spending too much time with the businesses." (Interview 8, 2015)

It is useful to note that not all organisations joined the Friends of ETS coalition for the same reasons. For example, in some cases, electricity utilities associated the idea of pushing for ETS reform with the ability to raise electricity prices in the face of falling demand and poorly performing generation investments. As a coalition participant put it:

"...it kind of completely undermines our entire argument"... "but, as long as the [*desired*] outcome is the same, then it doesn't really matter what everybody's intentions are." (Interview 30, 2016)

The breadth of the coalition led to some tension and constraints on what it could achieve. Although the coalition was ambivalent about the issue of a single target, some of the most powerful members were strongly in favour, shaping the membership and the policy preferences of the group:

This is where [*the coalition*] was boxed into a corner. So with the Friends of ETS, because we had a couple of companies who were the single target coalition...Statoil, E.ON; we couldn't talk about multiple targets, which meant we couldn't get the [*renewable energy*] people on board. (Interview 30, 2016)

Importantly, the approach to assembling the coalition was more organic than by design, with political judgment, staying power and trust seen to be more important qualities in members than comprehensive representation of a range of interests:

Lots of people like grand designs, you want to get vanguards from every single sector, [*but it is important to pick people that you*] trust, people who understand politics, who understand it's not a straight line. And that if things go a bit wobbly they're able to keep people together because things will always go wobbly. (Interview 30, 2016)

In some cases, conventional trade associations were able to team up with other groupings. For example, EuroACE, the energy efficiency industry association, played a full role in the Coalition for Energy Savings, alongside other groups. The European Wind Energy Association (EWEA) was a member of the Prince of Wales Corporate Leaders Group.

It should be noted that membership of a group does not automatically translate into influence over its position. In the same way that some businesses felt that BUSINESSEUROPE's policy positions did not reflect their interests (Interview 8, 2015), Shell, a big oil refiner and therefore energy-intensive consumer, for example, was unable to persuade the Prince of Wales Corporate Leaders Group to support a position in favour of greater carbon leakage provisions (Interview 18, 2015).

The most common way of aggregating the policy views of a broad group of actors with common interests takes the form of a trade association. This model and its federated variant have been familiar features of the Brussels political landscape since the 1950s (Greenwood 2011; Coen 2007; Coen and Richardson 2009). The largest, oldest and, on paper, the weightiest of these is BUSINESSEUROPE which claims to speak on behalf of 20 million businesses (BUSINESSEUROPE 2016). But, despite the impressive representation, BUSINESSEUROPE and the energy intensives' credibility with policymakers was often poor.

This appears to be for three reasons. Firstly, BUSINESSEUROPE and the energy-intensive industries had, during the ETS reform debate been seen to inflexibly resist the policy around which a consensus of other actors had developed, putting them, in intellectual terms, on the periphery of the discussion. Second, BUSINESSEUROPE, along with the energy-intensive industries, failed to reach out to a larger constituency during the ETS reform and 2030 debates and was perceived by many people, inside and outside the Commission, to speak on largely behalf of the energy-intensive industries (Harvey 2013). Finally, it was an active strategy of some self-described 'progressive' business actors to counter the sheer mass of BUSINESSEUROPE's representation by highlighting, wherever possible, that BUSINESSEUROPE's 'does not speak for business in Europe' (Interview 8, 2015).

Simply because BUSINESSEUROPE's ideas were treated with some scepticism did not mean that they were not an important presence in the debate. The opinion of BUSINESSEUROPE was still important to policymakers, especially those in political positions in the Commission such as Energy Commissioner Oettinger who took the concerns of heavy industry very seriously and served as a 'steer' to other lobbyists and campaigners as to the limits of a debate. This is primarily due to the sheer size of its constituency. To put the relative size of BUSINESSEUROPE into context, while it was able to claim that it spoke on behalf of companies that employ 120 million people, another trade association, EuroACE, the energy efficiency group was able to field an equivalent number of 350,000 employees (Interview 12, 2015).

The ECF's deployment of its substantial financial resources relies on building networks and coalitions. First, ECF's commitment to 'the cause, not the brand' means that they tend not to engage in active lobbying, instead contributing to the creation of resources which other actors can use. It starts with the identification of a:

...knowledge gap, as we call it, in terms of data and a technical understanding of the problem, we [*engage in*] some technical analysis. [*Which
becomes*] a bit of a reference point for the NGOs. So we can profession-
alise or improve their contribution that they can do. And we frame the
debate around what we think are the key levers and the key choices for the
debates. (Interview 11, 2015)

It was significant that ECF's modelling contributions from 2010 onwards
were drafted and reviewed by actors of its 'core reflection group' representing a wide range of interests. Since the assumptions and methodologies
used in the analysis were vetted and approved by the entire group, which
was carefully selected by ECF to represent as many interests as possible, the
results were more acceptable to policymakers (Interview 10, 2015).

Although this approach to team building was not without its risks.
Members of the group that did not necessarily share ECF's policy goals,
such as the gas industry, began to use the assumptions and analysis as
the basis of secondary analysis in support of their own policy positions
(Interview 11, 2015).

Intelligence Gathering

Information about what other actors are doing, how they are doing it,
and changes in the receptiveness of policymakers to certain ideas is seen
as an important strategic asset. As mentioned in Chapter 4, one goal
of lobbyists is to become the 'captains of information'. In the 2030
debate, teamwork and ad hoc coalition building played a distinct role in
the sharing and dissemination of intelligence. On joining a group, actors
gain access to the pooled informational resources of other members.
Different actors assessed the quality of intelligence differently. For example, for many in the business community, the intelligence gathered by
NGOs was particularly important:

Their [*ENGOs*] political intelligence and network are not to be underesti-
mated, certainly, which is why we try to coordinate with them as much as
possible. (Interview 23, 2015)

We all talk to each other and we all need each other, and you know,
sometimes we will share intelligence behind the scenes because we're
all trying to do the same thing to a certain degree even if it's nuanced.
(Interview 29, 2016)

Of course, we also talk with the more classical, environment NGOs like
Greenpeace, WWF and also the Coalition for Energy Savings to discuss, to

exchange, information and exchange intelligence on how things are evolving.... (Interview 5, 2015)

For members of the Friends of ETS coalition, access to information was an important reason for participating.

It [*the coalition*] is an exchange of intelligence, information. Shared meetings, contact details...it was really gathering information that we had from meetings to have a broad lobbying effort. (Interview 1, 2015)

Moreover, the coalition recognised that managing access to the group was, in effect, managing access to information:

One of the things [*we*] did was create an email address, and you've got to be on this email address because all of the information, all of the leaks, all of the intelligence is coming on that email list. [*We were*] getting leaks and sharing it with them before [*we*] shared it with anybody else. (Interview 30, 2016)

Within the Brussels policy community, people often move between organisations with an interest in the same policy area. Less often, they cross the fault-lines that exist within the community that we describe in Chapter 4. On those rare occasions, people find themselves playing the role of interlocutor between two camps. In one example, a former ENGO lobbyist moved into the electricity sector and while some former colleagues felt that:

'...now we can talk to them. We can talk to you and, therefore, to people we find it hard to talk to.' (Interview 8, 2015)

Division of Labour
Despite the relatively small number of *bonafide* policy experts in Brussels, EU decision-making processes involve a comparatively large number of people. The complexity of effectively engaging with the process means that, in many cases, actors seek to share the burden of work required in mounting a campaign:

"But I think the main driver [*of coalition building*] in the end is that it's a very big process and there really aren't enough bodies on the ground to cover it. We don't all speak technical languages, et cetera, etcetera, etcetera."... "Nobody can do it alone. (Interview 8, 2015)

A lobbyist for a large utility company that was a member of the 'Friends of ETS' coalition, when discussing the coalition's ability to engage in dynamic, fast-moving campaigning described how different members brought different strengths to the group:

> "Companies [*like us*] are not set up for campaigning in that way. We had experts in communications, in campaigns, in the coalition. Some members have more competencies in that area - campaigning by NGOs for example. We pool together our resources..." "it's about sharing the load." (Interview 1, 2015)

Allocating and managing resources was important for Friends of ETS as illustrated below:

> "'[*We asked*], hey, can we borrow your office? Can we get someone in? Can we use your facilities?' Knowing that if Shell sent an email off to people saying, 'Look, we're serious about doing something for the ETS'; people are going to turn up to it...whenever there's an official discussion, we have sessions beforehand; 'this is the message that we want to say, who is going to be speaking, who should be speaking', then different people start to influence the people who are setting the calendar for who's speaking and we make sure our guys are there." (Interview 30, 2016)

6.4.5   Leading by Example

The nature of climate and energy policy means that there was little scope for civil society actors to show the ideas in action or '*create working models of the proposed policy*' themselves (Mintrom and Norman 2009). Activity like proving the viability of new energy technology is often beyond the means of even the biggest players (Interview 25, 2015). Some, however, were able to point to examples of policy implementations outside Europe. The argument made by actors in favour of limiting climate ambition often hinged on painting Europe as a 'first-mover' which, should it adopt ambitious unilateral climate targets, would be detrimental to Europe's international competitiveness.

> The way to answer that as a lobbyist was to keep showing a map produced by the World Bank of what's going on with carbon market developments globally, keep saying, 'China. Brazil. California. Mexico. Korea. We're not alone.' (Interview 8, 2015)

Member states, however, were in a stronger position to demonstrate the workability of policy implementations. Several members had in place policies on climate change that were clearly more ambitious than others' were. The UK's Climate Change Act and Germany's ongoing Energiewende provided clear examples of similar policy at work. In the case of the Energiewende, though, it also provided an example of just how complex and challenging an economy-wide low-carbon transition could be.

6.5   CONCLUSION

This chapter presented evidence of the timing and nature of the policy window that opened, fluctuated and eventually closed in response to events in the politics and problem streams. The policy window available to advocates seeking to influence the 2030 targets was narrower and more difficult to navigate that the one which enabled the 20/20/20 agreement. However, from 2011, opportunities were available to shape what would become the 2030 framework. Not all actors experienced the same policy window, with opportunities opening at different times for different actors. Some actors, such as ECF, were able to put resources at risk earlier in the process than others through the creation of new information while most only engaged in the later stages of the discussion.

A wide range of actors attempted an equally wide range of entrepreneurial activity, although even the observed successes were partial. One of the most striking examples of attempted policy entrepreneurship was the creation of the Friends of ETS coalition in which a motley collection of actors were able to work together with a clear position in favour of EU-ETS reform. That upstream and downstream gas, wind, solar, utility, manufacturing, finance and other interests were able to agree on a common problem diagnosis and climate-friendly policy solution is somewhat impressive consequence of the social acuity, informational skills and team management of the coalition founder.

However, establishing a common position on EU-ETS reform was not easy. To accommodate the powerful gas and utility actors that were felt necessary to lend the coalition's message credibility, the group was unable to propose, promote or even openly discuss multiple targets, to which those actors were fiercely opposed. At the same time, the strategy to define the functioning of the EU-ETS as the primary problem facing policymakers compounded Friends of ETS' status as a de facto single target grouping. Actors that may have supported multiple targets in 2007 and were inclined to do so again in 2013/2014 were constrained in their ability to do so while simultaneously participating in the Friends of ETS coalition.

The result was a weakening of support for multiple targets during this period as an indirect result of the parallel negotiation of EU-ETS reform and the 2030 targets.

REFERENCES

Ackrill, R., & Kay, A. (2011). Multiple Streams in EU Policy-Making: The Case of the 2005 Sugar Reform. *Journal of European Public Policy, 18*(1), 72–89.

Ackrill, R., Kay, A., & Zahariadis, N. (2013). Ambiguity, Multiple Streams, and EU Policy. *Journal of European Public Policy, 20*(6), 871–887.

Beyers, J., et al. (2014). Let's Talk! On the Practice and Method of Interviewing Policy Experts. *Interest Groups & Advocacy, 3*, 174–187.

Beyers, J., & De Bruycker, I. (2017). Lobbying Makes (Strange) Bedfellows: Explaining the Formation and Composition of Lobbying Coalitions in EU Legislative Politics. *Political Studies*, 1–26. https://doi.org/10.1177/0032321717728408.

Blanqué, P., et al. (2015). *Investor Letter to Companies on EU Climate Policy Positions*. Available at: http://www.iigcc.org/publications/publication/investor-letter-to-companies-on-eu-climate-policy-positions. Accessed 19 July 2016.

Boasson, E. L., & Wettestad, J. (2013). *EU Climate Policy: Industry, Policy Interaction and External Environment*. Oxford: Ashgate.

Boscarino, J. E. (2009). Surfing for Problems: Advocacy Group Strategy in U.S. Forestry Policy. *Policy Studies Journal, 37*(3), 415–434.

BUSINESSEUROPE. (2016). *ASGroup—Our Partner Companies*. Available at: https://www.businesseurope.eu/about-us/asgroup-our-partner-companies. Accessed 4 Aug 2016.

Carter, N., & Jacobs, M. (2014). Explaining Radical Policy Change: The Case of Climate Change and Energy Policy Under the British Labour Government 2006–10. *Public Administration, 92*(1), 125–141.

CEZ Group. (2014). *CEOs Initiative*. CEZ.cz. Available at: http://www.cez.cz/en/cez-group/cez-group/public-affairs/ceos-initiative.html. Accessed 8 June 2016.

Coen, D. (2007). Empirical and Theoretical Studies in EU Lobbying. *Journal of European Public Policy, 14*(3), 333–345.

Coen, D., & Richardson, J. (2009). Institutionalizing and Managing Intermediation in the EU. In D. Coen & J. Richardson (Eds.), *Lobbying the European Union: Institutions, Actors, and Issues* (pp. 337–350). Oxford: Oxford University Press.

Crouch, D. (2014). *Lobbyist's Take on Renewables Causes It to Lose Friends*. FT. com. Available at: https://www.ft.com/content/9b05ad2a-5ab0-11e4-b449-00144feab7de. Accessed 1 Sept 2015.

E3G. (2014). *Driving Change and Opportunity Through Strategic Influencing*. Available at: https://www.e3g.org/docs/Driving_Strategic_Change_-_Westminster_Hub_March_2014.pdf. Accessed 25 July 2016.

ENGIE. (2014). *People Need to Accept the Concept of Paying for the Climate*. Available at: http://www.engie.com/en/group/opinions/groups-strategy/people-need-to-accept-the-concept-of-paying-for-the-climate/. Accessed 6 Sept 2016.

Eurelectric. (2009). *Power Choices: Pathways to Carbon-Neutral Electricity in Europe by 2050*. Available at: www.eurelectric.org/PowerChoices2050/. Accessed 4 Apr 2016.

European Commission. (2012). *Directive 2012/27/EU on Energy Efficiency*. Available at: http://eur-lex.europa.eu/legal-content/EN/TXT/PDF/?uri=CELEX:32012L0027&from=EN. Accessed 1 Dec 2016.

European Commission. (2013). *Green Paper: A 2030 Framework for Climate and Energy Policies*. Available at: http://ec.europa.eu/transparency/regdoc/rep/1/2013/EN/1-2013-169-EN-F1-1.pdf. Accessed 10 Feb 2014.

European Council. (2014). *European Council (23 and 24 October 2014) Conclusions on 2030 Climate and Energy Policy Framework*. Available at: http://www.consilium.europa.eu/uedocs/cms_data/docs/pressdata/en/ec/145397.pdf. Accessed 24 Oct 2014.

Fairbrass, J. (2013). Natural Allies or Strange Bedfellows? The Emerging Relations Between Business, Civil Society, and Government in Response to the Challenge of Climate Change. In E. Monaghan et al. (Eds.), *New Climate Alliances* (pp. 19–22). Birmingham: Centre for Low Carbon Futures.

Greek Presidency of the Council of the European Union. (2014). *Informal Meeting of Energy Ministers Athens, 15–16 May 2014 'Energy Security' Discussion Paper*. Available at: http://gr2014.eu/sites/default/files/DiscussionPaperonEnergySecurity.pdf. Accessed 27 Apr 2016.

Greenwood, J. (2011). *Interest Representation in the European Union*. Basingstoke: Palgrave Macmillan.

Harvey, F. (2013). Europe's Climate Chief Vows to Fight on to Save Emissions Trading Scheme. *The Guardian*. Available at: http://www.guardian.co.uk/environment/2013/apr/17/europe-climate-chief-vow-save-emissions-trading. Accessed 14 June 2016.

Herweg, N. (2016). Explaining European Agenda-Setting Using the Multiple Streams Framework: The Case of European Natural Gas Regulation. *Policy Sciences, 49*(1), 13–33.

Kingdon, J. W. (2010). *Agendas, Alternatives, and Public Policies* (2nd ed.). Harlow: Pearson.

Magritte Group. (2013). *Press Release: Heads of 12 Leading European Energy Companies Propose Concrete Measures to Rebuild Europe's Energy Policy.* Available at: https://www.engie.com/wp-content/uploads/2013/11/12CEO_VA_v4.pdf. Accessed 19 Feb 2014.

Mintrom, M., & Norman, P. (2009). Policy Entrepreneurship and Policy Change. *Policy Studies Journal, 37*(4), 649–667.

Monciardini, D. (2016). The 'Coalition of the Unlikely' Driving the EU Regulatory Process of Non-Financial Reporting. *Social and Environmental Accountability Journal, 36*(1), 76–89.

de Schoutheete, P. (2012). The European Council. In J. Peterson & M. Shackleton (Eds.), *The Institutions of the European Union* (pp. 43–67). Oxford: Oxford University Press.

UNFCCC. (2012). *Durban Climate Change Conference—November 2011.* Available at: http://unfccc.int/meetings/durban_nov_2011/meeting/6245/php/view/reports.php#c. Accessed 26 Feb 2018.

Warren, A. (2011). *Some Targets Are More Equal Than Others.* Available at: http://www.ukace.org/2011/05/some-targets-are-more-equal-than-others/. Accessed 10 June 2016.

Wettestad, J., & Jevnaker, T. (2016). *Rescuing EU Emissions Trading: The Climate Policy Flagship.* London: Palgrave Macmillan.

Zahariadis, N. (2007). Multiple Streams Framework: Structure, Prospects, Limitations. In P. A. Sabatier (Ed.), *Theories of the Policy Process* (pp. 65–92). Boulder, CO: Westview Press.

CHAPTER 7

Conclusions and Discussion

Abstract This chapter lays out the principal findings that flow from the preceding analysis and reflects on them in the light of the original objectives of the research. A range of findings is presented which largely complement existing explanations, although they also emphasise the potential for the idea of 'technology neutrality' to drive both cohesion and division in the policy community. This complicating factor is offered as an additional explanation for the nature of the 2030 EU climate and energy targets. Finally, the chapter assesses the book's contribution to existing theoretical and empirical literature.

Keywords Technology neutrality · Problem surfing · Spillover Strange bedfellows

7.1 INTRODUCTION

The EU 2030 targets that were decided on by the European Council in 2014 have become an integral part of the EU's 'Energy Union', an attempt to place energy cooperation at the heart of the European Project (Szulecki et al. 2016). The decision is likely to determine the key characteristics, as well as the likelihood of success, of Europe's low-carbon energy system transformation and play a role in setting the terms of the next phase of European integration. The targets represent a reduction

© The Author(s) 2018   127
O. Fitch-Roy and J. Fairbrass, *Negotiating the EU's 2030
Climate and Energy Framework*, Progressive Energy Policy,
https://doi.org/10.1007/978-3-319-90948-6_7

in ambition compared to the 2020 policy that was previously in place and a strong shift in favour of technology-neutral policy, a shift that the literature on energy transitions and transformations suggests will have far-reaching implications for the route, and prospects for success, of Europe's low-carbon energy transformation.

It has long been observed that the EU policymaking process is remarkably open to European interest group activity (Coen and Richardson 2009). Far from an exception, climate and energy policy is one of the most heavily lobbied portfolios in Brussels. Commercial, industrial and environmental interests were all impacted by decisions taken in relation to the 2030 targets and were actively represented in the policymaking process surrounding that policy decision. This book reports on a study that investigated the role played by the interaction of those interests and the policy process from which the European Commission's proposals for the 2030 framework emerged. This chapter concludes the book by restating the findings, assessing the study against the aims, objectives and questions set for it in the opening and situating the work within the existing literature.

7.2   DISCUSSION OF PRINCIPAL FINDINGS

In this section, the main findings are presented and discussed as a series of explanatory narratives.

7.2.1   A Change in the Political and Economic Context

The political and economic environment in which the 2030 targets were debated was radically different from those that shaped the 2020 package. The 2020 policy was implemented at almost the same moment that the global financial crisis triggered the Eurozone crisis and the economic backdrop to the debate of the 2030 policy was several years of recession and uncertainty. This timing had two important impacts. First, it had material effects on key EU energy and climate policies as well as changing people's perception of these policies. The recession and associated slump in industrial production were major contributors to the collapse of the price of emissions allowances in the EU-ETS. The prevailing economic conditions also highlighted the budgetary implications of the 2020 policy. In particular, the cost of support for renewable energy became a key reference point. Actors resistant to a strong renewable energy target sensed that policymakers would be receptive to arguments that emphasised cost efficiency.

Secondly, the recession and attendant crisis in the Eurozone was recognised as having contributed to a measurable decline in public trust in the EU's capacity to deal with Europe's problems. This growing scepticism was evident in a surge in support for populist, anti-EU political parties across Europe and was reflected in the election to the European Parliament of record numbers of anti-EU or anti-establishment candidates in the elections in spring 2014. At the same time, climate change had been slipping down the list of European citizens' chief concerns since COP 15 in Copenhagen in 2009, at which the EU's attempts to broker a global deal on climate change had failed. The issue of whether a policy solution could in some way resist the fragmentary forces being experienced by the Union was perceived by some civil society actors to be an important factor in persuading European policymakers, especially senior Commission figures, of their legitimacy.

Under these conditions, that the EU was less ambitious in its climate and energy policy in 2014 than it had been in 2007 is unsurprising. It is interesting to observe, however, how civil society actors chose to make use of these macro-scale political developments in their attempts to influence the shape of the 2030 framework.

7.2.2   Magritte Group: Utility CEOs Team Up

The Magritte Group's approach to lobbying was not developed until the debate about 2030 became one focused on targets in 2013. Building a team of companies with very similar interests was not especially difficult for Gerard Mestrallet, chief executive of GDF Suez (now Engie). The group was able, however, by dint of their combined size and status of the CEOs, to conduct a high-profile media campaign as well as secure meetings with influential figures in the Commission, Parliament and Member States. While the members participated in various coalitions, they were not instrumental in setting them up. The utility companies of the Magritte Group were able to connect events in the problem and political streams to their preferred policy of a single target. The group carefully defined the problem of unprofitable gas-fired electricity generation plants as an energy security problem, which they strongly implied was caused by the earlier 20/20/20 framework and its renewable energy target. They also conspicuously made the case that they would like to see 'more Europe' in solving some of these problems, a welcome sentiment in the European Commission facing low levels of trust in the EU by European citizens.

7.2.3   A Disorderly Renewable Energy Lobby

The renewable energy industry has been shown to have been a potent force in the 20/20/20 debate from 2005 to 2007 (Boasson and Wettestad 2013), but in 2013 and 2014 they were perceived by many other actors as weak and disorganised. Many of the problems stemmed from the collapse of the umbrella group, EREC, at an especially inopportune time. The demise of EREC meant that not only were the individual industries unable to coordinate among themselves but forming wider coalitions was much harder. Most notably, the renewable energy industry did not form any meaningful coalition with the energy efficiency sector, a natural ally in many ways, especially in the face of rapidly increasing acceptance of a single target approach towards the end of 2013. The renewables industries were also unable to effectively counter the assertion, from groups such as the Single Target Coalition and the Magritte Group that, not only had renewables contributed to the problems faced by the EU-ETS by undermining the price, but that they had also exacerbated the issue of energy security by undercutting existing gas-fired generation stations. They argue that renewables were making gas plants unprofitable, forcing their owners to withdraw them from service. The renewable energy industries individually pointed out the energy security benefits of domestically sourced renewables but, alone, none of the voices had the authority or credibility of BUSINESSEUROPE, the oil companies and the Magritte Group, all of whom at least partially framed renewables as a problem for energy security. The lack of coordination between the energy efficiency and renewables industry may have been partial because of the delay caused by the energy efficiency review required by the energy efficiency directive. Ultimately, the renewable energy lobby, which for many years had pushed the idea that renewable energy was domestic, and therefore secure, struggled to make a case in favour of strong targets for renewable energy, even at a time energy security fears peaked in mid-2014 with the Ukraine–Russia crisis.

7.2.4 Early Mover: European Climate Foundation

The European Climate Foundation's (ECF) ability to move very early and take the risk that the policy process would not evolve as it hoped was matched by its team-building skills. By selecting a group of actors for its 'core reflection group' that was balanced by design, including actors from all segments of the policy community added to the credibility of the modelling work and its findings. They put forward a distinctive diagnosis of the problem as one of increasing the interconnectedness of the European electricity system, improving energy efficiency and increasing the proportion of energy from renewables. But having taken advantage of a slim opening of the policy window around the few months following COP 15 in Copenhagen, ECF were far less prominent in the remainder of the debate, effectively 'passing the baton' to the disorganised renewable energy sectors which were unable to cement the ideas into a rapidly shifting agenda. Their strategy of opening up the process of problem definition to the wider group was also risky and saw their data used in several unsanctioned reports from both campaigning NGOs and the gas industry.

7.2.5 Discord Inside the European Commission

There were differences of opinion about the policy package in both the administrative and political levels of the European Commission. DG Clima had been separated from the DG Environment at the start of the Barroso II Commission in 2010, a move that had been controversial at the time with critics concerned that the new DG would focus on the ETS as the main or only instrument for tackling climate change, leaving little scope for other policy options. The institutional corollary to this concern was that, on many issues, such as the EU 2030 framework, DG Clima would hold joint drafting responsibility with another newly formed department, DG Energy, which had previously been merged with Transport. Within the two Directorates-general responsible for drafting and consulting on the 2030 framework, DG Clima, custodians of the EU emissions trading system, were concerned about the future of the landmark policy. Given the recent low price of allowances within the system and the general performance of the system since the policy's implementation in 2005, this concern was understandable and the DG was somewhat cautious about the possible interaction between policies such as renewables targets and the ETS. DG Energy, meanwhile, had units specifically created to manage technology-specific policy areas such as renewable energy (unit C1) and energy efficiency (unit C3) and its staff were thought to be more in favour of a multiple-target approach.

Meanwhile, as well as this 'horizontal' tension between DGs, there was a 'vertical' dimension. There was an increasing tendency for the Secretariat-General (SG) to take an interest in the day-to-day activity within the DGs. This centralisation of control over policy within the Commission was manifest when both DG Clima and DG Energy found themselves relieved of drafting responsibility for the 2030 framework and relegated to a role as advisors to a new unit within the SG, which was to draft the policy. The reasons for the SG taking over the drafting brief are not clear. Whether it was simply a means to overcome intractable dispute of the nature of the framework between DG Clima and DG Energy or whether it was part of an explicit strategy of senior Commission figures seeking to take control of what President Barroso called *'politically challenging'* policies in order to avoid conflict with the European Council cannot be discerned. What is evident is that the framework that emerged was a political compromise, firstly between Oettinger and a small number of Commissioners who supported a 35% target and the majority who supported a stronger target and secondly between what the Commission perceived were the disparate preferences of the member states.

It is also the case that the framework was more compromised than either DG would have liked. On climate ambition, the proposal was towards the lower end of the range supported by the analysis undertaken by DG Clima and the weak, non-binding targets on energy efficiency and renewables clashed with the belief within DG Energy that without specific policies in these areas, the emissions goals would be very hard to achieve.

7.2.6   *'Big' Doesn't Need to Be 'Smart'*

Both BUSINESSEUROPE and the direct representatives of the energy-intensive industries took a more-or-less conventional approach to their lobbying efforts. Their message was largely the same as it had been on all topics of climate and energy policy for years. Their inability or unwillingness to find common ground with other organisations and the rigidity with which they stuck to their demands led other lobbyists and service level Commission officials to view them with scepticism.

More senior Commission figures, such as President Barroso and Commissioner Oettinger, however, were more likely to take notice of the group's colossal membership and both have been shown to have been very receptive to a single target approach (Bürgin 2015). It may be that BUSINESSEUROPE did not employ innovative techniques and build coalitions simply because it was large enough and influential enough not to need any help in order to be heard. It could also be speculated that by making the threat that 'business', and by inference prosperity, may suffer due to climate policy, policymakers had little choice but to listen. BUSINESSEUROPE and the energy-intensive industries were able to point out that consumers and industries were under financial pressure and that relaxing climate ambition, perhaps by refraining from a deal until after the Paris climate conference, could ease that pressure. They were unable, however, to make convincing arguments about how to square their demands with the desire to maintain Europe's position as a climate leader. The core message that climate and energy policy contributes to higher energy costs was particularly salient for policymakers at a time of economic crisis and worries about the cost of energy. It would appear that business interests were able to make a case that was received by policymakers without recourse to entrepreneurial activity. This echoes the concerns about a policy process in which business interests wield more power simply due to the existence of an all-encompassing market system. As Lindblom puts it, it is:

> ...not human need and aspiration but of the market system [is] the fixed element in the light of which we think about policy. (Lindblom 1982, p. 333)

7.2.7   Softening Up: Good Ideas Take Time

Kingdon (2010, p. 127) refers to a process he describes as 'softening up' by which familiarity with a concept or idea within the policy community is increased over a period of time so that, when the time is right, the idea is generally accepted to be valid and does not come as a surprise. In 2014, trade association EuroACE was able to make use of the political space created by the escalating Russia–Ukraine energy dispute to make a strong argument in favour of an energy efficiency target directly to the energy ministers of member states. However, it was no accident that energy efficiency advocates could move quickly. They had prepared their arguments ahead of time and had engaged in several years of softening up. The idea that energy efficiency could reduce gas dependency and, therefore, play a role in improving energy security had been gently nurtured by EuroACE since it coined the term 'energy dependence day' in 2011.

The idea of technology neutrality and an ETS-only approach to climate and energy policy required little 'softening up' because it had been the among the most important and ambitious policy initiatives relating to the climate issue that the EU had embarked on in the previous decade. Familiarity with the trading system was widespread and it is highly unlikely that a Brussels lobbyist or other political actor could thrive without a solid understanding of the concept as well the details of its functioning and history.

7.2.8   Rethinking Spillover: Entrepreneurship Does Not Equal Power

The fact that there were two overlapping policy windows, one in which ETS reform was being legislated and the other in which the agenda was being set for the 2030 targets meant that it was always likely that some kind of spillover would occur. Within the ETS reform window, one of the most striking pieces of entrepreneurship appears to have taken place. Friends of ETS (FoETS), a coalition convened by a small ENGO managed to play an important role in securing the two reforms to the trading system, backloading and the market stability reserve, which it saw as crucial to increasing Europe's climate change mitigation efforts. FoETS worked tenaciously to build a large and diverse coalition of actors, which put pressure on policymakers to enact the changes to the ETS. The coalition appeared to be especially impressive since it spanned cleavages in the policy community between ENGOs, utilities, renewable energy producers, oil companies and non-energy businesses.

The coalition actively contributed to a perception that the ETS was in crisis, as evidenced by the very low price of allowances, and consistently emphasised the point that resolution of the crisis was the primary problem of EU climate and energy policy. In order to create this sense of crisis and a perception that ETS was *the* climate policy problem in need of a solution, FoETS went as far as to indicate that earlier policy decisions such as the 2020 renewable energy target, had contributed to the low prices in the ETS. The reasoning behind FoETS' agnostic stance on 'technology neutrality' was partly to enhance the ability build the coalition. By pragmatically adopting a position firmly in favour of the ETS as the central issue in climate policy, the coalition could appeal to actors such as Shell and RWE, long-standing critics of renewable energy targets. It was also very much in line with the thinking in the key Directorate-general on climate policy, DG Clima. DG Clima is, or at least some of the key figures leading the department were, deeply wedded to the idea of the EU-ETS as the 'cornerstone' of EU climate policy. The reason for the DGs existence was intimately bound to the maintenance and perfection of the trading system and its staff was judged more receptive to advocacy that shared that view. Indeed, the decision to pursue ETS reform by environmentalists was based on a perception that institutional support for maintaining the ETS was so strong—in spite the policy's manifest failure to effect change—that there was little point 'blowing against the wind' with alternative suggestions. The support of some strategically important members was also contingent on the group not taking a supportive line towards policies other than the ETS, especially energy targets.

While it was perceived by its members and others to have been successful in the ETS reform debate (Wettestad and Jevnaker 2016), the impact of the coalition and its activity on the 2030 debate was more ambiguous. As acknowledged by the participants, maintaining the cohesion of the group meant that on the key issue of single versus multiple targets it was silent or purposefully ambivalent. This suited some actors more than it suited others. For example, utilities like GDF Suez, E. On and CEZ and oil major Shell were able to remain members of the FoETS coalition while actively pursuing other coalitions lobbying hard for a single target. However, once the reforms to the ETS had been secured, the coalition continued to act as a platform for information sharing and social interaction among the members.

Pressing for ETS reform could be seen, in 2012, to be the best chance for environmental campaigners to make a tangible impact on the ambition of EU climate policy. The EU-ETS was perceived to be unmovable as the central plank of climate and energy policy and Europe's political and economic situation emphasised the significance of energy costs and the 'cost efficiency' of policy. This meant that a sizable group of businesses that saw themselves as 'progressive' on climate (or wanted others to do so) were able to participate in the FoETS coalition. Moreover, it is possible to argue, as the participants and many observers do, that FoETS was an extremely effective coalition for demonstrating support and arguing for ETS reform and, therefore, a successful environmental lobbying campaign. However, it is also possible to arrive at an alternative conclusion.

A coalition on the scale of FoETS, in the relatively small community of EU climate and energy specialists, was always likely to have a wider impact than simply on the EU-ETS reform debate. For some members, a pro- (reformed) ETS stance may have directly affected their approach to the 2030 debate. The European Wind Energy Association (EWEA), for example, was a FoETS member that had traditionally been a strong supporter of renewable energy targets. Participation in FoETS may have made making the case for renewable energy targets intellectually more challenging to sustain and there was certainly some suspicion among other actors that EWEA had come close to adopting a single target position.

Many of the 'progressive' members (i.e. excluding the utilities and oil companies) of the coalition may have been open, under other circumstances, to discussions about how best to promote the idea of multiple targets. Despite success for the environment in the ETS reform debate, the FoETS coalition contributed to a significant shift of the overall policy debate's centre of gravity towards technology neutrality and the idea of a single target.

Kingdon (2010) anticipated spillovers from one policy window to another, and it appears that spillover occurred between the ETS reform debate and the 2030 debate. What Kingdon did not apparently anticipate was the potential for an entrepreneur to establish a precedent or create networks in one window, which undermined or dampened the effects of their entrepreneurship in a second window.

7.2.9   A Fragmented Policy Community

Kingdon (2010, p. 118) suggests that the rate at which a new idea may take hold within a policy community is a function of how integrated is that community. The expectation is that the closer the integration of the policy community, the faster an idea can take hold and more fractured or fragmented policy communities resulting in the slower propagation of ideas.

During 2013 and 2014, the climate and energy policy community in Brussels was deeply fractured. The most energy-intensive and polluting industries as well as lobby-group, BUSINESSEUROPE had little or no constructive interaction with the rest of the community of businesses, think tanks, and ENGOs. This remainder was so closely integrated, however, that it may be more useful to think of the policy community as two separate communities.

While the community of energy-intensive industries had a strong and shared worldview, it did not interact a great deal with the other groups. The energy-intensive industries were both perceived, and perceived themselves to be, outside of the core of decision-making actors and tended to pursue other avenues to influence policy than persuasion and contributing to problem-solving within the policy community. Within the other community, however, while policy positions and opinions were varied, the basic underlying premise of the need to tackle climate change was shared and there was widespread and frequent interaction between its members. This community included many of the Commission officials working on drafting the proposals.

Despite the fragmentation into two, almost independent, communities, the fact that one of the communities was very closely integrated meant that during 2013, the idea of an ETS-only climate and energy policy, which had been on the fringes of the debate for some time, moved very fast into the mainstream of the community's thinking.

7.3   Revisiting the Objectives of the Book

As well as the explicit goal of exploring the role of interest groups in determining the nature of the energy targets in the 2030 framework, this study set out to meet two additional aims. These are the creation of a detailed impression of the climate and energy policy community in 2013 and 2014 and a description of the context of the debate about targets.

This book has described and analysed a wide range of factors. Personal beliefs, interpersonal relationships and trust have been shown to be significant, as have grand historical and political trends. But, at the heart of the book is the story of an idea. More accurately, it is a small part of a much larger story of an idea. The idea is 'technology neutrality' and its story stretches all the way back to Pigou's suggestion in the 1920s that putting a price on economic 'externalities' is a justifiable means of modifying the behaviour of businesses and individuals in a way that reduces or eliminates the externality (Pigou 1932).

This idea has long been contested within the fields of climate and energy policymaking. The debate about climate and energy policy is not simply a unidimensional 'pull and push' between low and high climate ambition (i.e. between achieving climate goals and maintaining European industry). The additional, crosscutting tension between technology neutral and more technology-specific approaches give rise to an inherently multidimensional and altogether more complex debate. Technology neutrality is invoked and manipulated in support of a wide range of interests. Utility companies worried about the rate of renewable energy growth, oil and gas actors concerned about a future for gas in the energy mix, energy-intensive consumers concerned about the cost of climate policy, and member states attempting to retain sovereignty over energy policy decisions all invoked the idea at some point.

This breadth of appeal meant that the idea of 'technology neutrality' defined the battleground on which the 2030 targets were debated. It structured the policy community in a way that significantly influenced the 2030 framework, ultimately setting an agenda for a more technology neutral and, as we learn from transitions studies, probably less effective decarbonisation policy for Europe. As set out in the introduction, compared to the 20/20/20 policy package, in which specific renewable technology targets were derived for each member state, the 2030 package with its weak European-level renewable energy and energy efficiency targets was a distinct shift of EU policy in favour of technology neutrality.

7.4   Contribution to Literature

This book contributes to three main areas of literature. First, it augments existing accounts of how the 2030 targets came to be. Second, it makes some more general observations in relation to the role of interest groups in EU governance. Third, it contributes to the ongoing debate about the application and refinement of the multiple streams approach (MSA).

The preceding account largely validates the findings of earlier studies of the 2030 targets. We concur with others that business interests, divisions within the Commission and a confused European political landscape all contributed to the negotiation of a 2030 package that is less transformational than the earlier package for 2020 (Bürgin 2015; Fuchs and Feldhoff 2016; Ydersbond 2016).

This book is also an empirical contribution to the growing stock of descriptive and evaluative case studies about interest representation and the role of interest groups in the multilevel European governance structure (Coen 2007; Kohler-Koch 1994; Greenwood 2011; Richardson and Coen 2009). It furnishes fresh empirical evidence that validates many of the observations in the literature about collective action by European interest groups, especially with regard to their rationales for creating ad hoc coalitions of interests or maintaining hierarchical governance structures. At the same time, this book suggests a greater role for *ideas* in the mediation of power between interest groups than has previously been acknowledged, as well as the potential for interest groups to act apparently against their own stated interests as they seek to create more stable, influential coalitions.

In particular, it becomes clear that the observed trend towards closer cooperation between business and environmental interests in the field of climate policy lobbying is apparent and that it is an important factor in agenda-setting (Fairbrass 2013; Boasson and Huitema 2017). However, this kind of partnership is not necessarily neutral, and the costs (such as, in this case, taking a more technology-neutral approach in order to satisfy partners from business) and benefits (such as enhanced credibility for business) of creating business/ENGO coalitions can be unevenly distributed in favour of businesses. The obvious mitigation for any interests engaged in negotiating policy positions with other interest groups is to exercise caution.

Finally, this volume is an application of the MSA, contributing to an improved understanding of the validity of the approach's assumptions and the applicability of the approach to the analysis of EU interest representation.

By tightening the definition of policy entrepreneurship in MSA (Mintrom and Norman 2009), and allowing for interest group activity across all three of the streams (Rozbicka and Spohr 2016), the book has shown that MSA can be successfully applied to a case study of EU climate and energy policy interest representation. It also confirms that the original scheme, in which interest groups were excluded from the problem and politics streams, provides an inadequate conceptualisation of the status of interest groups in the policy process.

Kingdon (2010) and later Zahariadis (2014) and others have developed the MSA to explain the agenda-setting process under assumed conditions of ambiguity, manifest as unclear goals, fluid participation and unclear or opaque organisational technology. During the debate about the 2030 targets, these conditions generally appear to have been present. The goals of some actors do seem to have been unclear. Differences of opinion within the Commission made it difficult to ascertain the overarching policy goals. For example, industrial development, climate mitigation, minimising energy costs to consumers, diversifying supply away from Russia were all argued to be 'most' important at some point. Arguably, the 'Trilemma' frame introduced in Sect. 3.3 (page 37) and often cited by the Commission is, itself, an acknowledgement that the goals of energy policy can be somewhat ambiguous. The goals and preferences of civil society actors were also fuzzy. It was not entirely clear, for example, whether some environmental groups wanted to increase the effectiveness of the EU's main climate policy instrument (the EU-ETS) or encourage the transformation of the energy system. While on the surface they are apparently complementary goals, manipulation of the idea of technology neutrality by interests in favour of a single, technology-neutral GHG target made it very hard to argue strongly for ETS reform while also making the case for the renewable energy and energy efficiency targets that the literature on transitions and transformations suggest will be effective.

Participation was fluid within parts of the decisions-making process. For example, by moving the drafting responsibility away from of DG Energy and DG Clima into a specialist unit within the SG, a new set of personnel was required to become familiar with the policy brief. Civil society actors concentrated their efforts on different parts of the process. ECF was more involved early on while the utility companies focussed on attempting to influence the agenda much later. The renewable energy lobby more-or-less departed from view as EREC collapsed during the process.

It is less easy to be certain that the 'organisational technology', to use Zahariadis' term, was unclear. All actors seemed to have understood the process by which the Commission was able to propose a framework and the European Council was able to endorse it as policy. Actors can hold opposing views, however, on the most important elements of the process. For example, some lobbyists emphasise intergovernmental bargaining as critical and therefore target the Council while others focus on the persuasion of Commission desk officers by civil society. The role of parliamentarians in the agenda-setting process also seemed to be a source of some confusion. Whether this diversity of preferred access points is evidence of actors 'venue-shopping' to make the most of their resources (Mazey and Richardson 2001) or a less-than-perfect understanding of the processes is not clear.

Another of the core assumptions of MSA was difficult to validate. That the three streams function independently of one another is central to the model's conception, although Kingdon himself had some reservations about how separate they really are and was concerned in particular about the separation of the problem and policy streams, acknowledging that there was a certain degree of relatedness (Kingdon 2010, p. 227). When using the MSA as a heuristic for organising and analysing research findings, the researcher must repeatedly answer the question 'to which stream does this observation belong?' In most cases, the answer is straightforward: the energy security crisis is clearly a 'problem' and a renewable energy target (for example) is clearly a 'policy'. However, some confusion arose around the topic of the EU-ETS.

The ETS is, very obviously, a policy of the EU. However, the ETS and its performance were considered a major problem for EU climate and energy policy. Deciding in which stream this 'problematic policy' should be handled posed something of a dilemma. Ultimately, the crisis in the ETS was treated by this book as a 'problem', largely based on how the issue was presented by members of the policy community. This decision was based on an assessment that since the 2030 targets were not (directly) about the ETS, it provided something of a background to the debates about targets, rather than a particular solution or element of the policy.

The ambiguity about whether the EU-ETS deserves to considered within the policy or the problem streams is problematic for MSA, which is built on the premise that the streams are separate at all times outside stream coupling. However, it should be borne in mind that the separation of the streams is a conceptual conceit rather than a description of the real world and the device appears to have enabled a useful and illuminating analysis. In general, Kingdon's (2010, p. 141) observation that problem and policies evolve independently and that when it comes to policy ideas, '*there is no new thing under the sun*' is well supported by the prominent role in the debate the idea of technology neutrality. Indeed, the idea bears all the hallmarks of '*a solution searching for a problem*' that characterises the 'garbage can' model on which MSA is built (Zahariadis 2008, p. 519; Kingdon 2010; Cohen et al. 1972, p. 2).

As argued by Spohr and Rozbicka (2016), the activity of interest groups and other civil society actors can be captured by the use of the MSA to analysis. Not only does this study support the assertion that civil society actors can and should be treated as full participants in all of the streams rather than merely an adjunct to the politics stream, it also shows that the framework can be used as an effective lens for a focused view on that particular part of each of the streams.

While together, all elements of MSA provided a valuable set of tools to collate and analyse information, some elements of MSA have proved themselves particularly useful. The MSA concept of spillover, for instance, enabled the study to conceptualise two interdependent processes (ETS reform and EU 2030), with actions taken in one having impacts in the other and the role of social processes within the policy community (attempts at policy entrepreneurship) in enabling such spillover.

In addition to a more tightly specified role of the policy entrepreneur, there are also several assumptions of the framework for which modifications or refinements may be suggested in light of the findings. In order to illustrate the importance of structural factors in determining how an when an entrepreneur should best deploy their resources, the metaphor of a surfer is often invoked (Kingdon 2010, p. 165; Boscarino 2009). It is held that an actor (or surfer) must be prepared for the fortuitous confluence of problems and/or politics that opens a policy window (or wave). Once the wave (opportunity) comes, the surfer must be ready to catch it. In the MSA metaphor the surfer, although they possess agency over which waves they attempt to catch and what they choose to do once one arrives, is destined to wait for serendipity to send a wave their way. The metaphor largely rings true. Individual actors cannot control large-scale political or problematic events like the Russia–Ukraine gas crisis or the rise of populist sentiment any more than a surfer can control the waves in the ocean.

However, the data suggest two qualifications for the metaphor. The first is about the assumed inability of actors to exert even modest control over the opening of policy windows: the breaking of waves. There was a certain inevitability that a policy window would open, due to the impending expiry of the existing climate and energy framework in 2020 and the upcoming Paris climate conference in December 2015. However, while these events made up the limits of the policy window, actors that moved early were able to impose some shape on the window. ECF, for example, in discussion with the Commission made an attempt to kick-start the debate as one about renewable energy and energy efficiency (and potential targets) with their modelling project at a point in 2009, when the shape of the debate to come was yet to take shape. ECF was unable to act to follow up their work in the light of later developments but they did contribute to the fact that targets were being discussed at all. It could be said that the ECF Roadmap 2050 project marked the opening of the usable policy window and the start of the debate about the 2030 targets.

The second qualification is the absence in the literature of much discussions of what may happen when catching waves or so-called problem surfing goes wrong. It is expected that pushing an idea into the wrong window or at the wrong moment simply leads to the proposal or campaign 'fizzling out' or being *'destroyed on the reef'* (Kingdon 2010, p. 171). However, the findings of this study show that something more damaging may occur to problem surfers. As discussed above, the Friends of the ETS coalition sought to make use of the peaking crisis in the ETS in order to make the case for ETS reform and did so with some success, primarily through the recruitment of business interests to the coalition. Through the mechanism of spillover into the adjacent 2030 targets policy window, its activity contributed to the rise up the agenda of the single target idea, reducing the scope for renewable energy and energy efficiency targets. This outcome could well be described as a 'wipeout'[1] for the surfer interested in the transformation of the EU energy system, rather than simply strengthening the EU-ETS.

References

Boasson, E. L., & Huitema, D. (2017). Climate Governance Entrepreneurship: Emerging Findings and a New Research Agenda. *Environment and Planning C: Politics and Space, 35*(8), 1343–1361.

Boasson, E. L., & Wettestad, J. (2013). *EU Climate Policy: Industry, Policy Interaction and External Environment*. Oxford: Ashgate.

Boscarino, J. E. (2009). Surfing for Problems: Advocacy Group Strategy in U.S. Forestry Policy. *Policy Studies Journal, 37*(3), 415–434.

Bürgin, A. (2015). National Binding Renewable Energy Targets for 2020, but Not for 2030 Anymore: Why the European Commission Developed from a Supporter to a Brakeman. *Journal of European Public Policy, 22*(5), 690–707.

Coen, D. (2007). Empirical and Theoretical Studies in EU Lobbying. *Journal of European Public Policy, 14*(3), 333–345.

Coen, D., & Richardson, J. (2009). Institutionalizing and Managing Intermediation in the EU. In D. Coen & J. Richardson (Eds.), *Lobbying the European Union: Institutions, Actors, and Issues* (pp. 337–350). Oxford: Oxford University Press.

---

[1] A wipeout is a fall from a surfboard. For an especially vivid description of the experience of surfing, see Finnegan (2015).

Cohen, M. D., March, J. G., & Olsen, J. P. (1972). A Garbage Can Model of Organizational Choice. *Administrative Science Quarterly, 17*(1), 1–25.

Fairbrass, J. (2013). Natural Allies or Strange Bedfellows? The Emerging Relations Between Business, Civil Society, and Government in Response to the Challenge of Climate Change. In E. Monaghan et al. (Eds.), *New Climate Alliances. Centre for Low Carbon Futures* (pp. 19–22). Birmingham: University of Birmingham.

Finnegan, W. (2015). *Barbarian Days: A Surfing Life.* London: Corsair.

Fuchs, D., & Feldhoff, B. (2016). Passing the Scepter, Not the Buck: Long Arms in EU Climate Politics. *Journal of Sustainable Development, 9*(6), 58.

Greenwood, J. (2011). *Interest Representation in the European Union.* Basingstoke: Palgrave Macmillan.

Kingdon, J. W. (2010). *Agendas, Alternatives, and Public Policies* (2nd ed.). Harlow: Pearson.

Kohler-Koch, B. (1994). Changing Patterns of Interest Intermediation in the European Union. *Government and Opposition, 29*(2), 166–180.

Lindblom, C. E. (1982). The Market as Prison. *The Journal of Politics, 44*(2), 324–336.

Mazey, S., & Richardson, J. (2001). Interest Groups and EU Policy-Making: Organisational Logic and Venue Shopping. In J. Richardson (Ed.), *European Union: Power and Policy-Making* (pp. 247–268). New York: Routledge.

Mintrom, M., & Norman, P. (2009). Policy Entrepreneurship and Policy Change. *Policy Studies Journal, 37*(4), 649–667.

Pigou, A. C. (1932). *The Economics of Welfare* (4th ed.). London: Macmillan.

Richardson, J., & Coen, D. (2009). *Lobbying the European Union: Institutions, Actors, and Issues.* Oxford: Oxford University Press.

Rozbicka, P., & Spohr, F. (2016). Interest Groups in Multiple Streams: Specifying Their Involvement in the Framework. *Policy Sciences, 49*(1), 55–69.

Szulecki, K., et al. (2016). Shaping the 'Energy Union': Between National Positions and Governance Innovation in EU Energy and Climate Policy. *Climate Policy, 16*(5), 548–567.

Wettestad, J., & Jevnaker, T. (2016). *Rescuing EU Emissions Trading: The Climate Policy Flagship.* London: Palgrave Macmillan.

Ydersbond, I. M. (2016). *Where Is Power Really Situated in the EU?* Oslo: Fridtjof Nansen Institute.

Zahariadis, N. (2008). Ambiguity and Choice in European Public Policy. *Journal of European Public Policy, 15*(4), 514–530.

Zahariadis, N. (2014). Ambiguity and Multiple Streams. In P. A. Sabatier & C. M. Weible (Eds.), *Theories of the Policy Process* (pp. 25–57). Boulder, CO: Westview Press.

LIST OF INTERVIEWS

Interview	Role	Organisation	Date of interview
1	Public affairs professional	European energy utility co.	16/04/15
2	Representatives X2	ENGO	09/04/15
3	Senior official	European Commission (SecGen)	20/11/15
4	Parliamentarian	European Parliament	07/05/15
5	Representative	Solar energy industry	03/08/15
6	Senior official	European Commission	10/07/15
7	Official	European Commission (SecGen)	21/10/15
8	Public affairs professional	Electricity industry	04/09/15
9	Public affairs professional	Oil producer	25/02/15
10	Senior official	European Commission (DG Energy)	21/09/15
11	Public affairs professional	ENGO	02/07/15
12	Public affairs professional	Energy efficiency industry	26/08/15
13	Public affairs professional	Renewable energy industry	15/04/15
14	Representative	NGO	08/09/15
15	Public affairs professional	Business advocacy	14/04/15
16	Public affairs professional	Gas industry	05/11/15
17	Senior official	European Commission (DG Energy)	16/04/15
18	Analyst	Electricity industry	17/04/15
19	Analyst	European think tank	06/11/15
20	Public affairs professional	Steel industry	28/07/15

(continued)

O. Fitch-Roy and J. Fairbrass, *Negotiating the EU's 2030 Climate and Energy Framework*, Progressive Energy Policy, https://doi.org/10.1007/978-3-319-90948-6

Interview	Role	Organisation	Date of interview
21	Public affairs professional	Energy efficiency industry	10/04/15
22	Representative	ENGO	26/02/15
23	Public affairs professional	Renewable energy industry	24/04/15
24	Public affairs professional	European energy utility co.	23/02/15
25	Public affairs professional	Oil producer	24/02/15
26	Public affairs professional X2	European energy utility co.	12/01/16
27	Public affairs professional X2	Coal industry	12/01/16
28	Senior official	European Commission (DG Energy)	12/01/16
29	Public affairs professional	Business advocacy	08/01/16
30	Representative	ENGO	13/01/16
31	Public affairs professional	Business advocacy, emissions trading	20/02/15
32	Representative	ENGO	29/04/16

BIBLIOGRAPHY

Ackrill, R., & Kay, A. (2011). Multiple Streams in EU Policy-Making: The Case of the 2005 Sugar Reform. *Journal of European Public Policy, 18*(1), 72–89.

Ackrill, R., Kay, A., & Zahariadis, N. (2013). Ambiguity, Multiple Streams, and EU Policy. *Journal of European Public Policy, 20*(6), 871–887.

Adelle, C., & Anderson, J. (2013). Lobby Groups. In A. Jordan & C. Adelle (Eds.), *Environmental Policy in the EU: Actors, Institutions and Processes*. London and New York: Routledge.

Andresen, T. (2014). *German Utilities Hammered in Market Favoring Renewables—Bloomberg Business*. Bloomberg.com. Available at: http://www.bloomberg.com/news/articles/2013-08-11/german-utilities-hammered-in-market-favoring-renewables. Accessed 1 Dec 2015.

Awerbuch, S. (2006). Portfolio-Based Electricity Generation Planning: Policy Implications for Renewables and Energy Security. *Mitigation and Adaptation Strategies for Global Change, 11*(3), 693–710.

Azar, C., & Sandén, B. A. (2011). The Elusive Quest for Technology-Neutral Policies. *Environmental Innovation and Societal Transitions, 1*(1), 135–139.

Babiker, M. H. (2005). Climate Change Policy, Market Structure, and Carbon Leakage. *Journal of International Economics, 65*(2), 421–445.

Balanyá, B., et al. (2003). *Europe Inc: Regional and Global Restructuring and the Rise of Corporate Power* (2nd ed.). London: Pluto Press.

Baumgartner, F. R., et al. (2009). *Lobbying and Policy Change: Who Wins, Who Loses, and Why*. Chicago and London: University of Chicago Press.

Baumgartner, F. R., & Jones, B. (1991). Agenda Dynamics and Policy Subsystems. *The Journal of Politics, 53*(4), 1044–1074.

© The Editor(s) (if applicable) and The Author(s) 2018   147
O. Fitch-Roy and J. Fairbrass, *Negotiating the EU's 2030 Climate and Energy Framework*, Progressive Energy Policy,
https://doi.org/10.1007/978-3-319-90948-6

Beckman, K. (2013). 'Progressive Energy Companies' Versus Magritte Group. Available at: http://www.energypost.eu/progressive-energy-companies-versus-margritte-group/. Accessed 1 Feb 2014.

Bendor, J., Moe, T. M., & Shotts, K. W. (2001). Recycling the Garbage Can: An Assessment of the Research Program. American Political Science Review, 95(1), 169–190.

Benson, D., & Russel, D. (2015, January). Patterns of EU Energy Policy Outputs: Incrementalism or Punctuated Equilibrium? West European Politics, 38(1), 185–205.

Beyers, J., et al. (2014). Let's Talk! On the Practice and Method of Interviewing Policy Experts. Interest Groups & Advocacy, 3, 174–187.

Beyers, J., & De Bruycker, I. (2017). Lobbying Makes (Strange) Bedfellows: Explaining the Formation and Composition of Lobbying Coalitions in EU Legislative Politics. Political Studies, 1–26. https://doi.org/10.1177/0032321717728408.

Birchfield, V. L. (2011). The Role of EU Institutions in Energy Policy Formation. In V. L. Birchfield & J. Duffield (Eds.), Toward a Common European Union Energy Policy (pp. 235–262). London and New York: Palgrave Macmillan.

Birkland, T. A. (2010). An Introduction to the Policy Process: Theories, Concepts and Models of Public Policy Making. Oxford: Routledge.

Blanqué, P., et al. (2015). Investor Letter to Companies on EU Climate Policy Positions. Available at: http://www.iigcc.org/publications/publication/investor-letter-to-companies-on-eu-climate-policy-positions. Accessed 19 July 2016.

BMWi. (2014). EEG 2014. Available at: http://www.bmwi.de/DE/Themen/Energie/Erneuerbare-Energien/eeg-2014.html. Accessed 31 July 2016.

Boasson, E. L., & Huitema, D. (2017). Climate Governance Entrepreneurship: Emerging Findings and a New Research Agenda. Environment and Planning C: Politics and Space, 35(8), 1343–1361.

Boasson, E. L., & Wettestad, J. (2013). EU Climate Policy: Industry, Policy Interaction and External Environment. Oxford: Ashgate.

Börzel, T. A. (1998). Organizing Babylon: On the Different Conceptions of Policy Networks. Public Administration, 76(2), 253–273.

Boscarino, J. E. (2009). Surfing for Problems: Advocacy Group Strategy in U.S. Forestry Policy. Policy Studies Journal, 37(3), 415–434.

Bouwen, P. (2002). Corporate Lobbying in the European Union: The Logic of Access. Journal of European Public Policy, 9(3), 365–390.

Bressanelli, E., Koop, C., & Reh, C. (2016). The Impact of Informalisation: Early Agreements and Voting Cohesion in the European Parliament. European Union Politics, 17(1), 91–113.

Buchan, D. (2009). *Energy and Climate Change: Europe at the Crossroads.* Oxford: Oxford University Press.

Buchan, D. (2013). *Why Europe's Energy and Climate Policies Are Coming Apart.* Available at: https://www.oxfordenergy.org/wpcms/wp-content/uploads/2013/07/SP-28.pdf. Accessed 20 Mar 2014.

Buchan, D., & Keay, M. (2014). *The EU's New Energy and Climate Goals for 2030: Under-Ambitious and Over-Bearing?* Available at: https://www.oxfordenergy.org/wpcms/wp-content/uploads/2014/01/The-EUs-new-energy-and-climate-goals-for-2030.pdf. Accessed 20 Mar 2014.

Bürgin, A. (2015). National Binding Renewable Energy Targets for 2020, but Not for 2030 Anymore: Why the European Commission Developed from a Supporter to a Brakeman. *Journal of European Public Policy, 22*(5), 690–707.

BUSINESSEUROPE. (2013). *2030 Green Paper Response.* Available at: https://www.businesseurope.eu/sites/buseur/files/media/imported/2013-00699-E.pdf. Accessed 19 Feb 2016.

BUSINESSEUROPE. (2016). *ASGroup—Our Partner Companies.* Available at: https://www.businesseurope.eu/about-us/asgroup-our-partner-companies. Accessed 4 Aug 2016.

Cairney, P., & Jones, M. D. (2016). Kingdon's Multiple Streams Approach: What Is the Empirical Impact of this Universal Theory? *Policy Studies Journal, 44*(1), 37–58.

Caldecott, B., & McDaniels, J. (2014). *Stranded Generation Assets: Implications for European Capacity Mechanisms, Energy Markets and Climate Policy Working Paper.* Available at: http://www.smithschool.ox.ac.uk/research/sustainable-finance/publications/Stranded-Generation-Assets.pdf. Accessed 10 Mar 2014.

Cameron, P. D. (2011). The EU and Energy Security: A Critical Review of the Legal Issues. In A. Antoniadis, R. Schütze, & E. Spaventa (Eds.), *The European Union and Global Emergencies: A Law and Policy Analysis* (pp. 125–166). London: Bloomsbury.

Cañizares, C., Rouco, L., & Andersson, G. (2009). Angle, Voltage and Frequency Stability. In A. Gomez-Exposito, A. J. Conejo, & C. Canizares (Eds.), *Electric Energy Systems: Analysis and Operation.* Boca Raton: CRC Press.

Carbon Brief. (2013). *Climate Rhetoric: What's an Energy Trilemma?* Available at: https://www.carbonbrief.org/climate-rhetoric-whats-an-energy-trilemma. Accessed 10 Oct 2016.

Carter, N., & Jacobs, M. (2014). Explaining Radical Policy Change: The Case of Climate Change and Energy Policy Under the British Labour Government 2006–10. *Public Administration, 92*(1), 125–141.

CEZ Group. (2014). *CEOs Initiative*. CEZ.cz. Available at: http://www.cez.cz/en/cez-group/cez-group/public-affairs/ceos-initiative.html. Accessed 8 June 2016.

Chazan, G., & Wiesmann, G. (2013). *Shale Gas Boom Sparks EU Coal Revival*. FT.com. Available at: https://www.ft.com/content/d41c2e8a-6c8d-11e2-953f-00144feab49a. Accessed 10 Oct 2016.

Cherp, A., et al. (2016, May). Comparing Electricity Transitions: A Historical Analysis of Nuclear, Wind and Solar Power in Germany and Japan. *Energy Policy, 101,* 612–628.

Cherp, A., et al. (2018, September 2017). Integrating Techno-economic, Socio-technical and Political Perspectives on National Energy Transitions: A Meta-theoretical Framework. *Energy Research and Social Science, 37,* 175–190.

Chester, L. (2010). Conceptualising Energy Security and Making Explicit Its Polysemic Nature. *Energy Policy, 38*(2), 887–895.

Chyong, C.-K., & Tcherneva, V. (2015). *Europe's Vulnerability on Russian Gas*. Available at: http://www.ecfr.eu/article/commentary_europes_vulnerability_on_russian_gas. Accessed 1 Nov 2016.

Coalition of Progressive Energy Companies. (2014). *Energy Companies Call for an Ambitious and Binding Renewables Target for 2030*. Available at: http://www.eneco.com/nl/~/media/cor/pdf/organisatie/030314coalitionletterrenewablestarget2030.ashx. Accessed 4 Apr 2015.

Cobb, R., & Elder, C. D. (1972). *Participation in American Politics: The Dynamics of Agenda Building*. Boston: Allyn and Bacon.

de Cock, C. (2010). *iLobby.eu: Survival Guide to EU Lobbying*. Delft: Eburon.

Coen, D. (1997). The Evolution of the Large Firm as a Political Actor in the European Union. *Journal of European Public Policy, 4*(1), 91–108.

Coen, D. (2007). Empirical and Theoretical Studies in EU Lobbying. *Journal of European Public Policy, 14*(3), 333–345.

Coen, D., & Richardson, J. (2009). Institutionalizing and Managing Intermediation in the EU. In D. Coen & J. Richardson (Eds.), *Lobbying the European Union: Institutions, Actors, and Issues* (pp. 337–350). Oxford: Oxford University Press.

Cohen, M. D., March, J. G., & Olsen, J. P. (1972). A Garbage Can Model of Organizational Choice. *Administrative Science Quarterly, 17*(1), 1–25.

Cornot-Gandolphe, S. (2015). *US Coal Exports: The Long Road to Asian Markets*. Available at: https://www.oxfordenergy.org/wpcms/wp-content/uploads/2015/03/CL-21.pdf. Accessed 28 Nov 2016.

Crouch, D. (2014). *Lobbyist's Take on Renewables Causes It to Lose Friends*. FT.com. Available at: https://www.ft.com/content/9b05ad2a-5ab0-11e4-b449-00144feab7de. Accessed 1 Sept 2015.

Crum, B. (2013). Saving the Euro at the Cost of Democracy? *Journal of Common Market Studies, 51*(4), 614–630.

Curtin, D. (2003). Private Interest Representation or Civil Society Deliberation? A Contemporary Dilemma for European Union Governance. *Social and Legal Studies, 12,* 55–75.

Dahl, R. A. (1978). Pluralism Revisited. *Comparative Politics, 10*(2), 191–203.

Dallos, G. (2014). *Locked in the Past: Why Europe's Big Energy Companies Fear Change.* Hamburg: Greenpeace.

Davey, E. (2013a). *Department of Energy and Climate Change Blog: Europe Must Stay Ambitious on Climate Change.* Available at: http://blog.decc.gov.uk/2013/05/28/europe-must-stay-ambitious-on-climate-change/. Accessed 4 Nov 2014.

Davey, E. (2013b). *Edward Davey Speech: Ambitious and Flexible—Europe's 2030 Framework for Emissions Reduction.* Available at: https://www.gov.uk/government/speeches/edward-davey-speech-ambitious-and-flexible-europes-2030-framework-for-emissions-reduction. Accessed 4 Nov 2014.

DECC. (2013). *2030 Green Paper Response.* Available at: https://www.gov.uk/government/uploads/system/uploads/attachment_data/file/210659/130703_response_for_publication.pdf. Accessed 10 Feb 2014.

Delreux, T., & Happaerts, S. (2016). *Environmental Policy and Politics in the European Union.* London: Palgrave Macmillan.

Dickel, R., et al. (2014). *Reducing European Dependence on Russian Gas: Distinguishing Natural Gas Security from Geopolitics.* Oxford: Oxford Institute for Energy Studies.

Dimitrov, R. S. (2010). Inside UN Climate Change Negotiations: The Copenhagen Conference. *Review of Policy Research, 27*(6), 795–821.

Dryzek, J., & Dunleavy, P. (2009). *Theories of the Democratic State.* Basingstoke: Palgrave Macmillan.

Duffield, J., & Birchfield, V. L. (2011). *Toward a Common European Union Energy: Problems, Progress, and Prospects.* Basingstoke: Palgrave Macmillan.

Dunlop, C. (2000). Epistemic Communities: A Reply to Toke. *Politics, 20*(3), 137–144.

Dupont, C. (2016). *Climate Policy Integration into EU Energy Policy: Progress and Prospects.* Oxford: Routledge.

Dür, A., & De Bièvre, D. (2007). The Question of Interest Group Influence. *Journal of Public Policy, 27*(1), 1–12.

Duscha, V., Held, A., & del Rio, P. (2016). An Economic Analysis of the Interactions Between Renewable Support and Other Climate and Energy Policies. *Energy & Environment, 28*(1–2), 11–33.

E3G. (2014). *Driving Change and Opportunity Through Strategic Influencing.* Available at: https://www.e3g.org/docs/Driving_Strategic_Change_-_Westminster_Hub_March_2014.pdf. Accessed 25 July 2016.

EDP. (2013). *2030 Green Paper Response*. Available at: https://crowdsourcing. simpolproject.eu/static/staticdata/gpc/consultations/edp_energias_de_portugal.pdf. Accessed 15 Apr 2016.

Eichhammer, W. (2013). *Analysis of a European Reference Target System for 2030. Energy Savings 2030: On the 2050 Pathway*. Available at: http://www. isi.fraunhofer.de/isi-wAssets/docs/x/de/publikationen/Fraunhofer-ISI_ ReferenceTargetSystemReport.pdf. Accessed 11 Feb 2016.

Eikeland, P. O. (2008). *EU Internal Energy Market Policy: New Dynamics in the Brussels Policy Game?* Available at: https://www.fni.no/getfile.php/132068/ Filer/Publikasjoner/FNI-R1408.pdf. Accessed 13 Feb 2014.

Eikeland, P. O. (2011). The Third Internal Energy Market Package: New Power Relations Among Member States, EU Institutions and Non-State Actors? *Journal of Common Market Studies, 49*(2), 243–263.

Eikeland, P. O. (2012). *EU Energy Policy Integration—Stakeholders, Institutions and Issue-Linkages*. Available at: https://www.fni.no/getfile.php/132050/ Filer/Publikasjoner/FNI-R1312.pdf. Accessed 13 Feb 2014.

Eising, R. (2002). Policy Learning in Embedded Negotiations: Explaining EU Electricity Liberalization Policy Learning in Embedded Negotiations: Explaining EU Electricity Liberalization. *International Organization, 56*(1), 85–120.

Elzen, B., Geels, F. W., & Green, K. (2004). *System Innovation and the Transition to Sustainability*. Cheltenham: Edward Elgar.

ENGIE. (2014). *People Need to Accept the Concept of Paying for the Climate*. Available at: http://www.engie.com/en/group/opinions/groups-strategy/ people-need-to-accept-the-concept-of-paying-for-the-climate/. Accessed 6 Sept 2016.

EPIA. (2013). *2030 Green Paper Response*. Available at: https://crowdsourcing. simpolproject.eu/static/staticdata/gpc/consultations/epia.pdf. Accessed 18 Apr 2016.

EREC. (2011). *45% by 2030: Towards a Truly Sustainable Energy System in the EU*. Brussels.

EREC. (2013). *Hat-Trick 2030 Renewable Energy Energy Efficiency Greenhouse Gas*. Brussels.

Euracoal. (2014). *Why Less Climate Ambition Would Deliver More for the EU*. Brussels.

Euracoal. (2015). *Climate Change*. euracoal.eu. Available at: https://euracoal. eu/coal/climate-change/. Accessed 12 Oct 2016.

Euractiv. (2013a). *Hedegaard: More 2030 Climate Targets Would Be 'Wise'*. Euractiv.com. Available at: http://www.euractiv.com/energy/hedegaard-2030-climate-targets-w-news-530979. Accessed 19 Apr 2016.

Euractiv. (2013b). *Oettinger Hails 'Wide Agreement' on 2030 Energy Targets, but Doubts Persist*. Euractiv.com. Available at: http://www.euractiv.com/

energy/2030-energy-target-doubts-oettin-news-530613. Accessed 4 May 2016.

Euractiv. (2014a). *Big EU Guns Fire for 'Crucial' 2030 Renewable Targets.* Euractiv.com. Available at: http://www.euractiv.com/energy/big-eu-guns-fire-crucial-2030-re-news-532608. Accessed 4 May 2016.

Euractiv. (2014b). *Denmark Signals Fight for Tougher 2030 Climate and Clean Energy Goals.* Euractiv.com. Available at: http://www.euractiv.com/energy/denmark-signals-fight-tougher-20-news-533025. Accessed 2 Feb 2016.

Euractiv. (2014c). *EU Sets Out 'Walk Now, Sprint Later' 2030 Clean Energy Vision.* Euractiv.com. Available at: http://www.euractiv.com/energy/eu-sets-walk-sprint-2030-clean-e-news-532960. Accessed 4 May 2016.

Euractiv. (2014d). *Germany Calls for Three 2030 Climate and Energy Targets.* Available at: http://www.euractiv.com/section/energy/news/germany-calls-for-three-2030-climate-and-energy-targets/. Accessed 4 May 2016.

Euractiv. (2014e). *Green MEP: Lobbyists Stopped Ambitious EU Energy Targets.* Euractiv.com. Available at: http://www.euractiv.com/sections/energy/green-mep-lobbyists-stopped-ambitious-eu-energy-targets-309867. Accessed 23 Feb 2015.

Euractiv. (2014f). *Green MEPs, NGOs Protest Commission's New 2030 Climate and Energy Targets.* Available at: http://www.euractiv.com/section/energy/video/green-meps-ngos-protest-commission-s-new-2030-climate-and-energy-targets/. Accessed 2 Apr 2016.

Euractiv. (2014g). *Member States' Positions on 2030 Climate and Energy Targets Revealed.* EurActiv.com. Available at: http://www.euractiv.com/sections/energy/member-states-positions-2030-climate-and-energy-targets-revealed-309279. Accessed 13 Apr 2016.

Euractiv. (2014h). *Oettinger Feels the Heat Over Climate Remarks.*EurActiv.com. *Available at: https://www.euractiv.com/section/energy/news/oettinger-feels-the-heat-over-climate-remarks/. Accessed 5 Apr 2016.*

Euractiv. (2014i). *Oettinger Rallies Opposition to 2030 CO2 Target.* Available at: https://www.euractiv.com/section/trade-society/news/oettinger-rallies-opposition-to-2030-co2-target/. Accessed 5 Apr 2016.

Euractiv. (2014j). *Poland Says It 'Won' at the EU Summit.* Euractiv. Available at: http://www.euractiv.com/sections/energy/poland-says-it-won-eu-summit-309494. Accessed 24 Nov 2014.

Eurelectric. (2009). *Power Choices: Pathways to Carbon-Neutral Electricity in Europe by 2050.* Available at: www.eurelectric.org/PowerChoices2050/. Accessed 4 Apr 2016.

Eurelectric. (2013). *2030 Green Paper Response.* Available at: https://www3.eurelectric.org/media/110882/eurelectric_-_2030_green_paper_consultation_response_-_final-2013-030-0486-01-e.pdf. Accessed 13 Apr 2016.

Eurelectric. (2014). *Power for a Competitive Europe: A Manifesto for a Balance, More Efficient European Energy Policy.* Available at: http://www.eurelectric. org/media/119468/manifesto_designed-2014-030-0083-01-e.pdf. Accessed 21 Apr 2016.

EuroACE. (2013). *EuroACE Position Paper on EU Post 2020 Climate and Energy Policy: EuroACE Supports a Binding 2030 Energy Efficiency Target.* Brussels.

EuroACE. (2014). *EuroACE Congratulates European Parliament for Its Economic Rationale in Calling for a Binding 40% Energy Efficiency Target and Sectoral Target for Buildings.* Brussels.

Eurogas. (2013). *The Eurogas 10 Point Plan to 2030.* Available at: http://www. eurogas.org/uploads/media/The_Eurogas_10-Point_Plan_to_2030.pdf. Accessed 14 July 2015.

European Climate Foundation. (2010a). *Newsletter Summer 2010.* Available at: http://www.europeanclimate.org/documents/ECF_Newsletter_ Summer2010.pdf. Accessed 31 Mar 2016.

European Climate Foundation. (2010b). *Roadmap 2050: A Practical Guide to a Prosperous, Low-Carbon Europe* (Vol. 2). Europe. Available at: http://www. roadmap2050.eu/. Accessed 31 Mar 2016.

European Climate Foundation. (2010c, April). *Roadmap 2050: A Practical Guide to a Prosperous, Low-Carbon Europe* (Vol. 1, p. 99). Available at: http://www.roadmap2050.eu/. Accessed 31 Mar 2016.

European Commission. (2006). *A European Strategy for Sustainable, Competitive and Secure Energy.* Available at: http://europa.eu/documents/comm/ green_papers/pdf/com2006_105_en.pdf. Accessed 14 Apr 2014.

European Commission. (2008). *Eurobarometer Special Report: Climate Change 2008.* Available at: http://ec.europa.eu/commfrontoffice/publicopinion/ archives/ebs/ebs_300_full_en.pdf. Accessed 3 May 2016.

European Commission. (2009a). *Commissioner Piebalgs Welcomes the Commitment of European Electricity Companies to Achieve a Carbon-Neutral Power Supply by 2050.* Available at: http://europa.eu/rapid/press-release_ IP-09-417_en.pdf. Accessed 31 March 2016.

European Commission. (2009b). *Directive 2009/72/EC Concerning Common Rules for the Internal Market in Electricity and Repealing Directive 2003/54/ EC.* Available at: http://eur-lex.europa.eu/legal-content/EN/TXT/ PDF/?uri=CELEX:32003L0054&from=EN. Accessed 13 Feb 2014.

European Commission. (2009c). *Eurobarometer Special Report: Climate Change 2009.* Available at: http://ec.europa.eu/commfrontoffice/publicopinion/ archives/ebs/ebs_313_en.pdf. Accessed 3 May 2016.

European Commission. (2011a). *A Roadmap for Moving to a Competitive Low Carbon Economy in 2050.* Available at: http://ec.europa.eu/clima/documen- tation/roadmap/docs/com_2011_112_en.pdf. Accessed 4 Apr 2016.

European Commission. (2011b). *Energy Roadmap 2050.* Available at: https://ec.europa.eu/energy/sites/ener/files/documents/2012_energy_roadmap_2050_en_0.pdf. Accessed 14 July 2015.

European Commission. (2011c). *Energy Roadmap 2050 Impact Assessment, Part 2/2: Accompanying the Document Energy Roadmap 2050.* Available at: https://ec.europa.eu/energy/sites/ener/files/documents/sec_2011_1565_part2.pdf. Accessed 4 Apr 2016.

European Commission. (2011d). *Eurobarometer Special Report: Climate Change 2011.* Available at: http://ec.europa.eu/commfrontoffice/publicopinion/archives/ebs/ebs_372_en.pdf. Accessed 3 May 2016.

European Commission. (2012). *Directive 2012/27/EU on Energy Efficiency.* Available at: http://eur-lex.europa.eu/legal-content/EN/TXT/PDF/?uri=CELEX:32012L0027&from=EN. Accessed 1 Dec 2016.

European Commission. (2013a). Draft Commission Staff Working Document Impact Assessment for a 2030 Climate and Energy Policy Framework.

European Commission. (2013b). *Europeans, the European Union and the Crisis: Standard Eurobarometer 79 Spring 2013.* Available at: http://ec.europa.eu/commfrontoffice/publicopinion/archives/eb/eb79/eb79_first_en.pdf. Accessed 4 May 2016.

European Commission. (2013c). *Green Paper: A 2030 Framework for Climate and Energy Policies.* Available at: http://ec.europa.eu/transparency/regdoc/rep/1/2013/EN/1-2013-169-EN-F1-1.pdf. Accessed 10 Feb 2014.

European Commission. (2013d). *Member States' Competitiveness Performance and Implementation of EU Industrial Policy.* Available at: http://ec.europa.eu/DocsRoom/documents/110/attachments/1/translations/en/renditions/native. Accessed 17 Feb 2016.

European Commission. (2013e). *Report on the Implementation, Results and Overall Assessment of the 2013 European Year of Citizens.* Available at: http://europa.eu/citizens-2013/sites/default/files/content/document/COM_2014_687_F1_REPORT_FROM_COMMISSION_EN_V4_P1_789180.PDF. Accessed 6 May 2016.

European Commission. (2013f). *Without a Strong Industrial Base, Europe's Economy Cannot Prosper.* Press release available at: http://europa.eu/rapid/press-release_IP-13-862_en.pdf. Accessed 28 Nov 2016.

European Commission. (2014a). *A Policy Framework for Climate and Energy in the Period from 2020 to 2030.* Available at: http://eur-lex.europa.eu/legal-content/EN/TXT/PDF/?uri=CELEX:52014DC0015&from=EN. Accessed 4 Feb 2014.

European Commission. (2014b). *Energy Efficiency and Its Contribution to Energy Security and the 2030 Framework for Climate and Energy Policy.* Available at: http://eur-lex.europa.eu/legal-content/EN/TXT/?uri=CELEX:52014DC0520. Accessed 20 Apr 2016.

European Commission. (2014c). *Energy Prices and Costs in Europe.* Available at: https://ec.europa.eu/energy/sites/ener/files/documents/20140122_communication_energy_prices.pdf. Accessed 7 Feb 2014.

European Commission. (2014d). *Eurobarometer Special Report: Climate Change 2014.* Available at: http://ec.europa.eu/commfrontoffice/publicopinion/archives/ebs/ebs_409_en.pdf. Accessed 3 May 2016.

European Commission. (2014e). *Impact Assessment—A Policy Framework for Climate and Energy in the Period from 2020 up to 2030.* Available at: http://ec.europa.eu/smart-regulation/impact/ia_carried_out/docs/ia_2014/swd_2014_0015_en.pdf. Accessed 13 July 2015.

European Commission. (2014f). *Minutes of the 2072nd Meeting of the Commission Held in Brussels (Berlaymont) on Wednesday 22 January 2014.* Available at: http://ec.europa.eu/transparency/regdoc/rep/10061/2014/EN/10061-2014-2072-EN-F1-1.Pdf. Accessed 20 Apr 2016.

European Commission. (2014g). *Proposal for the Establishment and Operation of a Market Stability Reserve for the Union Greenhouse Gas Emission Trading Scheme and Amending Directive 2003/87/EC.* Available at: http://eur-lex.europa.eu/legal-content/EN/TXT/PDF/?uri=CELEX:52014PC0020&from=en. Accessed 24 Nov 2015.

European Commission. (2015a). *Connecting Power Markets to Deliver Security of Supply, Market Integration and the Large-Scale Uptake of Renewables.* Available at: http://europa.eu/rapid/press-release_MEMO-15-4486_en.htm. Accessed 19 May 2016.

European Commission. (2015b). *Eurobarometer Special Report: Climate Change 2015.* Available at: http://data.europa.eu/euodp/en/data/dataset/S2060_83_4_435_ENG. Accessed 3 May 2016.

European Commission. (2017). *Proposed Regulation on the Governance of the Energy Union.* Available at: http://eur-lex.europa.eu/resource.html?uri=cellar:f9f04518-b7dc-11e6-9e3c-01aa75ed71a1.0001.02/DOC_1&format=PDF. Accessed 22 Feb 2018.

European Council. (2009). *29/30 October 2009: Conclusions.* Available at: http://www.consilium.europa.eu/uedocs/cms_data/docs/pressdata/en/ec/110889.pdf. Accessed 31 Mar 2016.

European Council. (2014). *European Council (23 and 24 October 2014) Conclusions on 2030 Climate and Energy Policy Framework.* Available at: http://www.consilium.europa.eu/uedocs/cms_data/docs/pressdata/en/ec/145397.pdf. Accessed 24 Oct 2014.

European Environment Agency. (2014). *Trends and Projections in Europe 2014: Tracking Progress Towards Europe's Climate and Energy Targets For 2020.* Available at: https://www.eea.europa.eu/publications/trends-and-projections-in-europe-2014/at_download/file. Accessed 28 Oct 2014.

European Environment Agency. (2016). *Annual European Union Greenhouse Gas Inventory 1990–2012 and Inventory Report 2014.* Available at: http://www.eea.europa.eu/publications/european-union-greenhouse-gas-inventory-2014. Accessed 22 Feb 2018.

European Gas Advocacy Forum. (2011). *Making the Green Journey Work.* Brussels.

European Parliament. (2014a). *Decision of the European Parliament and of the Council Concerning the Establishment and Operation of a Market Stability Reserve for the Union Greenhouse Gas Emission Trading Scheme and Amending Directive 2003/87/EC.* Brussels.

European Parliament. (2014b). *European Elections 2014.* Available at: http://www.europarl.europa.eu/us/en/elections_2014.html;jsessionid=AB-D017EA7C44EE440340530246C59FAA. Accessed 3 June 2014.

European Parliament. (2014c). *Report on a 2030 Framework for Climate and Energy Policies.* Available at: http://www.europarl.europa.eu/sides/getDoc.do?pubRef=-//EP//NONSGML+REPORT+A7-2014-0047+0+DOC+PDF+V0//EN. Accessed 31 Jan 2016.

European Parliament. (2017). *Minutes: Committee on Industry, Research and Energy, Meeting of 27 November 2017, 15.00–18.30, and 28 November 2017, 9.00–12.30 and 14.30–18.30.* Available at: http://www.europarl.europa.eu/sides/getDoc.do?type=COMPARL&reference=PE-615.215&format=PD-F&language=EN&secondRef=01. Accessed 22 Feb 2018.

Eurostat. (2014). *Energy Balance Sheets 2011–2012.* Available at: http://ec.europa.eu/eurostat/documents/3217494/5785109/KS-EN-14-001-EN.PDF/16c0ac97-7dd6-4694-b22d-e77a36cb4e86. Accessed 10 Feb 2016.

Eurostat. (2016). *Primary Production of Energy by Resource.* Available at: http://ec.europa.eu/eurostat/tgm/table.do?tab=table&init=1&language=en&pcode=ten00076&plugin=1. Accessed 22 Feb 2018.

EWEA. (2013). *2030 Green Paper Response.* Available at: https://crowdsourcing.simpolproject.eu/static/staticdata/gpc/consultations/ewea.pdf. Accessed 18 Apr 2016.

Fairbrass, J. (2013). Natural Allies or Strange Bedfellows? The Emerging Relations Between Business, Civil Society, and Government in Response to the Challenge of Climate Change. In E. Monaghan et al. (Eds.), *New Climate Alliances* (pp. 19–22). Leeds: Centre for Low Carbon Futures.

Falkner, R. (2014). Global Environmental Politics and Energy: Mapping the Research Agenda. *Energy Research & Social Science, 1,* 188–197.

Finnegan, W. (2015). *Barbarian Days: A Surfing Life.* London: Corsair.

Fitch-Roy, O. W. (2016). An Offshore Wind Union? Diversity and Convergence in European Offshore Wind Governance. *Climate Policy, 16*(5), 586–605.

Friends of the Earth. (2013). *Comments on the EC Green Paper 'A 2030 Framework for Climate and Energy Policies'.* Brussels.

Fuchs, D., & Feldhoff, B. (2016). Passing the Scepter, Not the Buck: Long Arms in EU Climate Politics. *Journal of Sustainable Development, 9*(6), 58.

Fürstenwerth, D., Pescia, D., & Litz, P. (2015). *The Integration Costs of Wind and Solar Power.* Agora Energiewende. Available at: https://www.agora-energiewende.de/fileadmin/Projekte/2014/integrationskosten-wind-pv/Agora_Integration_Cost_Wind_PV_web.pdf. Accessed 18 Feb 2016.

Gardner, P., Fitch-Roy, O. W., & Platt, R. (2012). *Beyond the Bluster: Why Wind Power Is an Effective Technology.* London: IPPR.

Garside, B. (2014). *Ministers from 7 EU Nations Call for Binding Energy Saving Goal.* Reuters.com. Available at: http://uk.reuters.com/article/eu-energy-efficiency-idUKL5N0OY4KQ20140617. Accessed 22 Apr 2016.

GDF Suez. (2013). *Consultation on a 2030 Framework for Climate and Energy Policies GDF SUEZ Answer.* Available at: http://ec.europa.eu/energy/consultations/doc/com_2013_0169_green_paper_2030_en.pdf. Accessed 19 Feb 2016.

Geels, F. W. (2002). Technological Transitions as Evolutionary Reconfiguration Processes: A Multi-Level Perspective and a Case-Study. *Research Policy, 31*(8–9), 1257–1274.

Geiger, A. (2012). *EU Lobbying Handbook* (2nd ed.). Brussels: Helios Media.

Government of Denmark. (2013). *2030 Green Paper Response.* Copenhagen.

Government of Poland. (2013). *2030 Green Paper Response.* Brussels.

Government of the Czech Republic. (2013). *2030 Green Paper Response.* Brussels.

Gray, V., & Lowery, D. (1996). A Niche Theory of Interest Representation. *The Journal of Politics, 58*(1), 91–111.

Greek Presidency of the Council of the European Union. (2014). *Informal Meeting of Energy Ministers Athens, 15–16 May 2014 'Energy Security' Discussion Paper.* Available at: http://gr2014.eu/sites/default/files/DiscussionPaperonEnergySecurity.pdf. Accessed 27 Apr 2016.

Greenwood, J. (2011). *Interest Representation in the European Union.* Basingstoke: Palgrave Macmillan.

Greenwood, J., & Aspinwall, M. D. (1998). *Collective Action in the European Union: Interests and the New Politics of Associability.* London and New York: Routledge.

Groen, L., Niemann, A., & Oberthür, S. (2012). The EU as a Global Leader? *The Copenhagen and Cancun UN Climate Change Negotiations, 8*(2), 173–191.

GSTEC. (2014). *Belgium: European Renewable Energy Council (EREC) Is History.* Available at: http://www.gstec.org/content/belgium-european-renewable-energy-council-erec-history. Accessed 15 Mar 2016.

Haas, E. B. (1958). *The Uniting of Europe: Political, Social and Economic Forces, 1950–1957*. Stanford, CA: Stanford University Press (Reprint, Notre Dame, IN: University of Notre Dame Press, 2003).

Haas, P. M. (1992). Epistemic Communities and International Policy Coordination. *International Organization, 46*(1), 1–35.

Harvey, F. (2013). Europe's Climate Chief Vows to Fight on to Save Emissions Trading Scheme. *The Guardian*. Available at: http://www.guardian.co.uk/environment/2013/apr/17/europe-climate-chief-vow-save-emissions-trading. Accessed 14 June 2016.

Helm, D. (2005). *European Energy Policy: Securing Supplies and Meeting the Challenge of Climate Change*. Oxford.

Herweg, N. (2015). Against All Odds: The Liberalisation of the European Natural Gas Market—A Multiple Streams Perspective. In K. S. Jale Tosun & S. Schmitt (Eds.), *Energy Policy Making in the EU: Building the Agenda* (pp. 87–105). London: Springer.

Herweg, N. (2016). Explaining European Agenda-Setting Using the Multiple Streams Framework: The Case of European Natural Gas Regulation. *Policy Sciences, 49*(1), 13–33.

Herweg, N., Huß, C., & Zohlnhöfer, R. (2015). Straightening the Three Streams: Theorising Extensions of the Multiple Streams Framework. *European Journal of Political Research, 54*(3), 435–449.

Herweg, N., & Zahariadis, N. (2018). The Multiple Streams Approach. In N. Zahariadis & L. Buonanno (Eds.), *The Routledge Handbook of European Public Policy* (pp. 32–42). Oxford: Routledge.

Herweg, N., Zahariadis, N., & Zohlnhöfer, R. (2018). The Multiple Streams Framework: Foundations, Refinements and Empirical Applications. In C. M. Weible & P. Sabatier (Eds.), *Theories of the Policy Process*. Boulder, CO: Westview Press.

Hoffmann, S. (1966). Obstinate or Obsolete? The Fate of the Nation-State and The Case of Western Europe. *Daedalus, 95*(3), 862–915.

Hone, D. (2015). *Putting the Genie Back: Why Carbon Pricing Matters*. Whitefox.

Howlett, M., Mcconnell, A., & Perl, A. (2017). Moving Policy Theory Forward: Connecting Multiple Stream and Advocacy Coalition Frameworks to Policy Cycle Models of Analysis. *Australian Journal of Public Administration, 76*(1), 65–79.

Hurrelmann, A. (2007). European Democracy, the 'Permissive Consensus' and the Collapse of the EU Constitution. *European Law Journal, 13*(3), 343–359.

IEA. (2016). *Gas Trade Flow in Europe*. Available at: https://www.iea.org/gtf/. Accessed 11 Apr 2016.

IFIEC. (2014). *Manifesto: Europe's Manufacturing Industry CEOs Call Upon Heads of State to Streamline 2030 Strategy Towards Growth and Jobs*. Brussels.

IFIEC Europe. (2013). *2030 Green Paper Response*. Brussels.

IPCC. (2014). *Climate Change 2014 Synthesis Report Summary Chapter for Policymakers*. Available at: https://www.ipcc.ch/pdf/assessment-report/ar5/syr/AR5_SYR_FINAL_SPM.pdf. Accessed 15 Aug 2016.

Janowska, K. (2011). Poland's Climate Change Policy Struggle. In R. Wurzel & J. Connelly (Eds.), *The European Union as a Leader in International Climate Change Politics*. Abingdon: Routledge.

Jones, B. D., & Baumgartner, F. R. (2005). *The Politics of Attention: How Government Prioritizes Problems*. Chicago: University of Chicago Press.

Jones, M. D., et al. (2016). A River Runs Through It: A Multiple Streams Meta-Review. *Policy Studies Journal, 44*(1), 13–36.

Joyce, A. (2014). *Speech to the Informal Energy Ministers Meetings—Athens—16th May 2014: Financing of Energy Efficiency Measures*. Athens.

Juergens, I., Barreiro-Hurlé, J., & Vasa, A. (2013). Identifying Carbon Leakage Sectors in the EU ETS and Implications of Results. *Climate Policy, 13*(1), 89–109.

Kanellakis, M., Martinopoulos, G., & Zachariadis, T. (2013). European Energy Policy—A Review. *Energy Policy, 62*, 1020–1030.

Karnitschnig, M. (2015). *Germany's Green Power Meltdown*. politico.eu. Available at: https://www.politico.eu/article/germanys-green-power-meltdown/. Accessed 22 Feb 2018.

Kern, F., & Howlett, M. (2009). Implementing Transition Management as Policy Reforms: A Case Study of the Dutch Energy Sector. *Policy Sciences, 42*(4), 391–408.

Kingdon, J. W. (2010). *Agendas, Alternatives, and Public Policies* (2nd ed.). Harlow: Pearson.

Kitzing, L., Mitchell, C., & Morthorst, P. E. (2012). Renewable Energy Policies in Europe: Converging or Diverging? *Energy Policy, 51*, 192–201.

Klüver, H. (2011). *Lobbying in Coalitions: Interest Group Influence on European Union Policy-Making*. Available at: https://www.nuffield.ox.ac.uk/politics/papers/2011/HeikeKluever_workingpaper_2011_04.pdf. Accessed 15 Aug 2016.

Klüver, H., Mahoney, C., & Opper, M. (2015). Framing in Context: How Interest Groups Employ Framing to Lobby the European Commission. *Journal of European Public Policy, 22*(4), 481–498.

Koch, N., & Mama, H. B. (2016). *European Climate Policy and Industrial Relocation: Evidence from German Multinational Firms*. Available at: https://papers.ssrn.com/sol3/Delivery.cfm/SSRN_ID2868283_code1302307.pdf?abstractid=2868283&mirid=1. Accessed 5 Dec 2016.

Kohler-Koch, B. (1994). Changing Patterns of Interest Intermediation in the European Union. *Government and Opposition, 29*(2), 166–180.

Kohler-Koch, B., & Eising, R. (2003). *Transformation of Governance in the European Union*. London and New York: Routledge.

LaFarge. (2013). *2030 Green Paper Response*. Brussels.

Laing, T., et al. (2013). *Assessing the Effectiveness of the EU Emissions Trading System* (CCCEP Working Paper No. 126).

Lane, P. R. (2012). The European Sovereign Debt Crisis. *Journal of Economic Perspectives, 26*(3), 49–68.

Lapavitsas, C., et al. (2010). Eurozone Crisis: Beggar Thyself and Thy Neighbour. *Journal of Balkan and Near Eastern Studies, 12*(4), 321–373.

Lindberg, L. N., & Scheingold, S. A. (1970). *Europe's Would-Be Polity: Patterns of Change in the European Community*. Eaglewood Cliffs: Prentice-Hall.

Lindblom, C. E. (1982). The Market as Prison. *The Journal of Politics, 44*(2), 324–336.

Long, T., & Loerinczi, L. (2009). NGOs as Gatekeepers: A Green Vision. In D. Coen & J. Richardson (Eds.), *Lobbying the European Union: Institutions, Actors, and Issues* (pp. 169–188). Oxford: Oxford University Press.

Lorenzoni, I., & Benson, D. (2014a). Radical Institutional Change in Environmental Governance: Explaining the Origins of the UK Climate Change Act 2008 Through Discursive and Streams Perspectives. *Global Environmental Change, 29*, 10–21.

Lorenzoni, I., & Benson, D. (2014b). Radical Institutional Change in Environmental Governance: Explaining the Origins of the UK Climate Change Act 2008 Through Discursive and Streams Perspectives. *Global Environmental Change, 29*, 10–21.

Łoskot-Strachota, A., & Zachmann, G. (2014). *Rebalancing the EU-Russia-Ukraine Gas Relationship*. Available at: https://www.econstor.eu/dspace/bitstream/10419/106321/1/812740270.pdf. Accessed 11 Feb 2016.

Lowery, D., & Gray, V. (1998). The Dominance of Institutions in Interest Representation: A Test of Seven Explanations. *American Journal of Political Science, 42*(1), 231–255.

Lowery, D., & Gray, V. (2004). A Neopluralist Perspective on Research on Organized Interests. *Political Research Quarterly, 57*(1), 164–175.

Lowery, D., & Gray, V. (2005). Sisyphus Meets the Borg. *Journal of Theoretical Politics, 17*(1), 41–74.

Lucas, N. J. D. (1977). *Energy and the European Communities*. London: Europa Publications.

Lund, H., & Mathiesen, B. V. (2009). Energy System Analysis of 100% Renewable Energy Systems—The Case of Denmark in Years 2030 and 2050. *Energy, 34*(5), 524–531.

Machiavelli, N. (1532). *The Prince*. London: Penguin (Reprint, Chicago: University of Chicago Press, 2010).

Magritte Group. (2013). *Press Release: Heads of 12 Leading European Energy Companies Propose Concrete Measures to Rebuild Europe's Energy Policy.* Available at: https://www.engie.com/wp-content/uploads/2013/11/12CEO_VA_v4.pdf. Accessed 19 Feb 2014.

Mahoney, C. (2007). Lobbying Success in the United States and the European Union. *Journal of Public Policy, 27*(1), 35–56.

Majone, G. (1994). The Rise of the Regulatory State in Europe. *West European Politics, 17*(3), 77–101.

Marcu, A., et al. (2013). *Carbon Leakage: An Overview.* Available at: https://www.ceps.eu/system/files/SpecialReportNo79CarbonLeakage_0.pdf. Accessed 11 Feb 2016.

Marks, G., & Hooghe, L. (2009). A Postfunctionalist Theory of European Integration: From Permissive Consensus to Constraining Dissensus. *British Journal of Political Science, 39*(1), 1–23.

Matlary, J. H. (1998). *Energy Policy in the European Union.* Basingstoke: Palgrave Macmillan.

Mazey, S., & Richardson, J. (2001). Interest Groups and EU Policy-Making: Organisational Logic and Venue Shopping. In J. Richardson (Ed.), *European Union: Power and Policy-Making* (pp. 247–268). New York: Routledge.

McCool, D. (1998). The Subsystem Family of Concepts: A Critique and a Proposal. *Political Research Quarterly, 51*(2), 551–570.

McFarland, A. S. (2007). Neopluralism. *Annual Review of Political Science, 10*(1), 45–66.

McGowan, F. (2011). The UK and EU Energy Policy: From Awkward Partner to Active Protagonist? In V. L. Birchfield & J. Duffield (Eds.), *Toward a Common European Union Energy Policy* (pp. 187–213). New York: Palgrave Macmillan.

Meadowcroft, J. (2009). What About the Politics? Sustainable Development, Transition Management, and Long Term Energy Transitions. *Policy Sciences, 42*(4), 323–340.

Meadowcroft, J. (2011). Engaging with the Politics of Sustainability Transitions. *Environmental Innovation and Societal Transitions, 1*(1), 70–75.

Mintrom, M., & Norman, P. (2009). Policy Entrepreneurship and Policy Change. *Policy Studies Journal, 37*(4), 649–667.

Mitchell, C., Bauknecht, D., & Connor, P. M. (2006). Effectiveness Through Risk Reduction: A Comparison of the Renewable Obligation in England and Wales and the Feed-in System in Germany. *Energy Policy, 34*(3), 297–305.

Mitchell, C., & Watson, J. (2013). Introduction: Conceptualising Energy Security. In C. Mitchell, J. Watson, & J. Britton (Eds.), *New Challenges in Energy Security: The UK in a Multipolar World* (pp. 1–21). Basingstoke: Palgrave Macmillan.

Monaghan, A. (2005). *Russian Oil and EU Energy Security*. Conflict. Available at: https://www.files.ethz.ch/isn/96125/05_Nov.pdf. Accessed 19 Feb 2016.

Monciardini, D. (2016). The 'Coalition of the Unlikely' Driving the EU Regulatory Process of Non-Financial Reporting. *Social and Environmental Accountability Journal, 36*(1), 76–89.

Moravcsik, A. (1998). *The Choice for Europe: Social Purpose and State Power from Messina to Maastricht* (1st ed.). London: Routledge.

Mucciaroni, G. (1992). The Garbage Can Model & the Study of Policy Making: A Critique. *Polity, 24*(3), 459–482.

Neslen, A. (2015). *Shell Lobbied to Undermine EU Renewables Targets, Documents Reveal*. Available at: http://www.theguardian.com/environment/2015/apr/27/shell-lobbied-to-undermine-eu-renewables-targets-documents-reveal. Accessed 27 Apr 2015.

Neslen, A. (2016). *EU Dropped Climate Policies After BP Threat of Oil Industry 'Exodus'*. Available at: http://www.theguardian.com/environment/2016/apr/20/eu-dropped-climate-policies-after-bp-threat-oil-industry-exodus. Accessed 22 Apr 2016.

Oberthür, S. (2011). The European Union's Performance in the International Climate Change Regime. *Journal of European Integration, 33,* 667–682.

Oberthür, S., & Roche Kelly, C. (2008). EU Leadership in International Climate Policy: Achievements and Challenges. *The International Spectator, 43*(3), 35–50.

Palmer, J. (2010). Stopping the Unstoppable? A Discursive-Institutionalist Analysis of Renewable Transport Fuel Policy. *Environment and Planning C: Government and Policy, 28*(6), 992–1010.

Pateman, C. (1970). *Participation and Democratic Theory*. Cambridge: Cambridge University Press.

Pigou, A. C. (1932). *The Economics of Welfare* (4th ed.). London: Macmillan.

Platts. (2013). *German Coal-Fired Power Rises Above 50% in First-Half 2013 Generation Mix—Electric Power | Platts News Article & Story*. Available at: http://www.platts.com/latest-news/electric-power/london/german-coal-fired-power-rises-above-50-in-first-26089429. Accessed 18 Feb 2016.

Princen, S. (2007). Agenda-Setting in the European Union: A Theoretical Exploration and Agenda for Research. *Journal of European Public Policy, 14*(1), 21–38.

Princen, S. (2018). Agenda-Setting and Framing in Europe. *The Palgrave Handbook of Public Administration and Management in Europe* (pp. 535–551). London, UK: Palgrave Macmillan.

Rawcliffe, P. (1995). Making Inroads: Transport Policy and the British Environmental Movement. *Environment, 37*(3), 16–36.

Rawcliffe, P. (1998). *Environmental Pressure Groups in Transition*. Manchester: Manchester University Press.

van Renssen, S. (2014a). *Climate Policy Bumps into Competitiveness in Europe*. Energypost.eu. Available at: http://www.energypost.eu/climate-policy-bumps-competitiveness-europe/. Accessed 13 July 2016.

van Renssen, S. (2014b). *Split Emerges in the Commission Over Energy-Efficiency Measures*. Politico. Available at: http://www.politico.eu/article/split-emerges-in-the-commission-over-energy-efficiency-measures/. Accessed 8 May 2014.

Reuters. (2014). *Portugal Could Block EU Climate Deal Over Connection Target*. Available at: http://www.reuters.com/article/eu-summit-climatechange-portugal-idUSL6N0SH1KI20141022. Accessed 18 May 2016.

Richardson, J. (2005). Policy-Making in the EU: Interests, Ideas and Garbage Cans of Primeval Soup. In J. Richardson (Ed.), *European Union: Power and Policy-Making* (pp. 3–30). New York: Routledge.

Richardson, J., & Coen, D. (2009). *Lobbying the European Union: Institutions, Actors, and Issues*. Oxford: Oxford University Press.

Ringel, M. (2006). Fostering the Use of Renewable Energies in the European Union: The Race Between Feed-in Tariffs and Green Certificates. *Renewable Energy, 31*(1), 1–17.

del Río, P., & Mir-Artigues, P. (2012). Support for Solar PV Deployment in Spain: Some Policy Lessons. *Renewable and Sustainable Energy Reviews, 16*(8), 5557–5566.

Rochefort, D. A., & Cobb, R. W. (1994). *The Politics of Problem Definition: Shaping the Policy Agenda*. Lawrence: University Press of Kansas.

Rosamond, B. (2000). *Theories of European Integration*. Basingstoke and New York: Palgrave Macmillan.

Rowlands, I. H. (2005). The European Directive on Renewable Electricity: Conflicts and Compromises. *Energy Policy, 33*(8), 965–974.

Rozbicka, P., & Spohr, F. (2016). Interest Groups in Multiple Streams: Specifying Their Involvement in the Framework. *Policy Sciences, 49*(1), 55–69.

Sabatier, P. A. (1998, March). The Advocacy Coalition Framework: Revisions and Relevance for Europe. *Journal of European Public Policy, 5,* 98–130.

Sabatier, P. A. (2007). *Theories of the Policy Process* (2nd ed.). Boulder, CO: Westview Press.

Sabatier, P. A., & Jenkins-Smith, H. C. (1993). *Policy Change and Learning: An Advocacy Coalition Approach*. Boulder, CO: Westview Press.

Sabatier, P. A., & Jenkins-Smith, H. C. (1999). The Advocacy Coalition Framework. In P. A. Sabatier (Ed.), *Theories of the Policy Process* (pp. 117–166). Boulder, CO: Westview Press.

Sandbag. (2013). *Consultation Response—2030 Energy and Climate Framework Green Paper*. Available at: https://crowdsourcing.simpolproject.eu/static/staticdata/gpc/consultations/sandbag.pdf. Accessed 30 Nov 2015.

Saurugger, S. (2008). Interest Groups and Democracy in the European Union. *West European Politics, 31*(6), 1274–1291.

van Schendelen, M. P. C. M. (2013). *The Art of Lobbying the EU: More Machiavelli in Brussels*. Amsterdam: Amsterdam University Press.

Schmitter, P. C. (1974). Still the Century of Corporatism? *The Review of Politics, 36*(1), 85–131.

Schmitter, P. C. (1977). Modes of Interest Intermediation and Models of Societal Change in Western Europe. *Comparative Political Studies, 10*(1), 7–38.

de Schoutheete, P. (2012). The European Council. In J. Peterson & M. Shackleton (Eds.), *The Institutions of the European Union* (pp. 43–67). Oxford: Oxford University Press.

Schuman, R. (1950). *Schuman Declaration*. Available at: https://europa.eu/european-union/about-eu/symbols/europe-day/schuman-declaration_en. Accessed 31 Oct 2016.

Sensfuß, F., Ragwitz, M., & Genoese, M. (2008). The Merit-Order Effect: A Detailed Analysis of the Price Effect of Renewable Electricity Generation on Spot Market Prices in Germany. *Energy Policy, 36*(8), 3076–3084.

Skjærseth, J. B. (2013). *Unpacking the EU Climate and Energy Package: Causes, Content and Consequences*. Available at: https://www.fni.no/getfile.php/131681/Filer/Publikasjoner/FNI-R0213.pdf. Accessed 16 Jan 2015.

Skjærseth, J. B. (2016). Linking EU Climate and Energy Policies: Policy-Making, Implementation and Reform. *International Environmental Agreements: Politics, Law and Economics, 16*(4), 509–523.

Skovgaard, J. (2013). The Limits of Entrapment: The Negotiations on EU Reduction Targets, 2007–11. *Journal of Common Market Studies, 51*(6), 1141–1157.

Solomon, B. D., & Krishna, K. (2011). The Coming Sustainable Energy Transition: History, Strategies, and Outlook. *Energy Policy, 39*(11), 7422–7431.

Spiegel, P., & Carnegy, H. (2014). *Anti EU Parties Celebrate Election Success*. FT.com. Available at: https://next.ft.com/content/783e39b4-e4af-11e3-9b2b-00144feabdc0. Accessed 5 Nov 2015.

Statoil ASA. (2013). *2030 Green Paper Response*. Brussels.

Van de Steeg, M. (2006). Does a Public Sphere Esist in the EU? An Analysis of the Content of the Debate on the Haider Case. *European Journal of Political Research, 45*, 609–634.

Stevens, P. (2012). *The 'Shale Gas Revolution': Developments and Changes*. London: Chatham House.

Stirling, A. (2010). Keep It Complex. *Nature, 468*(7327), 1029–1031.

Stirling, A. (2011). Pluralising Progress: From Integrative Transitions to Transformative Diversity. *Environmental Innovation and Societal Transitions, 1*(1), 82–88.

Stirling, A. (2014). Transforming Power: Social Science and the Politics of Energy Choices. *Energy Research and Social Science, 1,* 83–95.

Szulecki, K., et al. (2016). Shaping the 'Energy Union': Between National Positions and Governance Innovation in EU Energy and Climate Policy. *Climate Policy, 16*(5), 548–567.

The Coalition for Energy Savings. (2013). *A Binding Energy Savings Target for 2030: The Cornerstone for Mutually Supporting Climate and Energy Policies.* Available at: https://www.eurima.org/uploads/ModuleXtender/Publications/105/20131011_Coalition_position_on_2030.pdf. Accessed 7 Apr 2016.

The Economist. (2013). European Utilities: How to Lose Half a Trillion Euros. *The Economist.* Available at: http://www.economist.com/news/briefing/21587782-europes-electricity-providers-face-existential-threat-how-lose-half-trillion-euros. Accessed 16 Apr 2014.

The Economist. (2014). *The Eurosceptic Union.* Economist.com. Available at: http://www.economist.com/news/europe/21603034-impact-rise-anti-establishment-parties-europe-and-abroad-eurosceptic-union. Accessed 5 Nov 2016.

The Key Stakeholders Alliance for ETS Review. (2007). *Lowering Production is No Benefit for the Environment, Says European Industry.* Brussels.

The Prince of Wales's EU Corporate Leaders Group. (2013). *2030 Green Paper Response.* Cambridge: University of Cambridge.

The Royal Society. (2017). *Climate Updates: What Have We Learnt Since the IPCC 5th Assessment Report?* London: The Royal Society.

Torreblanca, J. I., & Leonard, M. (2013). *The Continent-Wide Rise of Euroscepticism.* London: ECFR.

Tosun, J., Biesenbender, S., & Schulze, K. (2015). *Energy Policy Making in the EU: Building the Agenda.* London: Springer.

UNFCCC. (2012). *Durban Climate Change Conference—November 2011.* Available at: http://unfccc.int/meetings/durban_nov_2011/meeting/6245/php/view/reports.php#c. Accessed 26 Feb 2018.

Valentine, S. V. (2011). Emerging Symbiosis: Renewable Energy and Energy Security. *Renewable and Sustainable Energy Reviews, 15*(9), 4572–4578.

Vattenfall. (2013). *2030 Green Paper Response.* Brussels.

Verbong, G., & Geels, F. W. (2007). The Ongoing Energy Transition: Lessons from a Socio-technical, Multi-level Analysis of the Dutch Electricity System (1960–2004). *Energy Policy, 35*(2), 1025–1037.

Votewatch Europe. (2015). *Who Holds the Power in the New European Parliament? And Why?* Available at: http://60811b39eee4e42e277a-72b421883bb5b133f34e068afdd7cb11.r29.cf3.rackcdn.com/2015/02/VoteWatch_template_web.pdf. Accessed 10 Jan 2017.

Warleigh, A. (2006). Making Citizens from the Market? NGOs and the Representation of Interests. In R. Bellamy, D. Castiglione, & J. Shaw (Eds.), *Making European Citizens: Civic Inclusion in a Transnational Context* (pp. 118–132). London: Palgrave Macmillan.

Warren, A. (2011). *Some Targets Are More Equal Than Others.* Available at: http://www.ukace.org/2011/05/some-targets-are-more-equal-than-others/. Accessed 10 June 2016.

Wathelet, M., et al. (2014, June 17). Letter Calling for a Proposal on a Binding Energy Efficiency Target for 2030.

Weible, C. M., & Sabatier, P. A. (2017). *Theories of the Policy Process* (4th ed.). New York: Routledge.

Weible, C. M., & Schlager, E. (2016). The Multiple Streams Approach at the Theoretical and Empirical Crossroads: An Introduction to a Special Issue. *Policy Studies Journal, 44*(1), 5–12.

Weiss, J. (2014). *Solar Energy Support in Germany: A Closer Look.* Washington, DC: Brattle.

Wettestad, J., Eikeland, P. O., & Nilsson, M. (2012). EU Climate and Energy Policy: A Hesitant Supranational Turn? *Global Environmental Politics, 12*(2), 67–86.

Wettestad, J., & Jevnaker, T. (2016). *Rescuing EU Emissions Trading: The Climate Policy Flagship.* London: Palgrave Macmillan.

Weyman-Jones, T. G. (1986). *Energy in Europe: Issues and Policies.* London and New York: Methuen.

Winkel, G., & Leipold, S. (2016). Demolishing Dikes: Multiple Streams and Policy Discourse Analysis. *Policy Studies Journal, 44*(1), 108–129.

Winkler, J., Magosch, M., & Ragwitz, M. (2018). Effectiveness and Efficiency of Auctions for Supporting Renewable Electricity—What Can We Learn from Recent Experiences? *Renewable Energy, 119,* 473–489.

Woll, C. (2007). Leading the Dance? Power and Political Resources of Business Lobbyists. *Journal of Public Policy, 27*(1), 57.

World Energy Council. (2015). *World Energy Trilemma Priority Actions on Climate Change and How to Balance the Trilemma.* Available at: http://www.worldenergy.org/wp-content/uploads/2015/05/2015-World-Energy-Trilemma-Priority-actions-on-climate-change-and-how-to-balance-the-trilemma.pdf. Accessed 18 Feb 2016.

WWF. (2012). *Re-energising Europe, Cutting Energy Related Emissions the Right Way.* Available at: http://awsassets.panda.org/downloads/cutting_energy_related_emissions_the_right_way_.pdf. Accessed 5 Apr 2016.

Ydersbond, I. M. (2016). *Where Is Power Really Situated in the EU?* Oslo: Fridtjof Nansen Institute.

Zachmann, G. (2015). *When Will the EU Switch Away from Coal?* Available at: http://bruegel.org/2015/12/when-will-the-eu-switch-away-from-coal/. Accessed 28 Nov 2016.

Zahariadis, N. (2007). Multiple Streams Framework: Structure, Prospects, Limitations. In P. A. Sabatier (Ed.), *Theories of the Policy Process* (pp. 65–92). Boulder, CO: Westview Press.

Zahariadis, N. (2008). Ambiguity and Choice in European Public Policy. *Journal of European Public Policy, 15*(4), 514–530.

Zahariadis, N. (2014). Ambiguity and Multiple Streams. In P. A. Sabatier & C. M. Weible (Eds.), *Theories of the Policy Process* (pp. 25–57). Boulder, CO: Westview Press.

INDEX

© The Editor(s) (if applicable) and The Author(s) 2018
O. Fitch-Roy and J. Fairbrass, *Negotiating the EU's 2030 Climate and Energy Framework*, Progressive Energy Policy,
https://doi.org/10.1007/978-3-319-90948-6

Printed by Printforce, the Netherlands